T0137032

Urban Informatics Using Mobile Network Data

Santi Phithakkitnukoon

Urban Informatics Using Mobile Network Data

Travel Behavior Research Perspectives

 Springer

Santi Phithakkitnukoon (iD)
Department of Computer
Engineering, Faculty of Engineering
Chiang Mai University
Chiang Mai, Thailand

ISBN 978-981-19-6716-0   ISBN 978-981-19-6714-6   (eBook)
https://doi.org/10.1007/978-981-19-6714-6

This Springer imprint is published by the registered company Springer Nature Singapore Pte Ltd.
The registered company address is: 152 Beach Road, #21-01/04 Gateway East, Singapore 189721,
Singapore

Preface

Urban informatics has emerged as an exciting and promising field requiring many interdisciplinary perspectives that involve computation as a central part by which methods and models are developed to generate a deeper understanding of city's functioning. The field of urban informatics is developing rapidly in the light of ubiquitous computing and sensor technologies. Due to its various potential benefits such as proactive urban planning, intelligent transportation systems, and traffic management, mobility studies have become central to urban informatics. A large part of urban informatics involves the development of data mining and machine learning algorithms for extracting and characterizing mobility patterns from heterogenous urban data sources. Opportunistic sensing data sources have recently become valuable to urban informatics and travel behavior research, especially mobile network data as it provides a longitudinal real-life behavior of a great amount of people. This book therefore discusses the role of mobile network data in urban informatics, particularly how the mobile network data is utilized in the mobility context, where approaches, models, and systems are developed for understanding travel behavior. The objectives of this book are thus to evaluate the extent to which mobile network data reflects travel behavior and to develop guidelines on how to best use such data to understand and model travel behavior. To achieve these objectives, the book attempts to evaluate the strengths and weaknesses of this data source for urban informatics and its applicability to the development and implementation of travel behavior models through a series of our research studies.

Traditionally, survey-based information is used as an input for travel demand models that predict future travel behavior and transportation needs. A survey-based approach is however costly and time-consuming, and hence its information can be dated and limited to a particular region. Mobile network data thus emerges as a promising alternative data source that is massive in both cross-sectional and longitudinal perspectives, and one that provides both broader geographic coverage of travelers and long-term travel behavior observation. The two most common types of travel demand model that have played an essential role in managing and planning for transportation systems are four-step models and activity-based models. The book's chapters are structured on the basis of these travel demand models in order to provide researchers and practitioners with an understanding of urban informatics and the important role that mobile network data plays in advancing the state-of-the-art from the perspectives of travel behavior research.

Chapter 1 introduces the key concepts and framework of urban informatics, discusses the use of mobile network data in urban informatics, and provides an outline of the book's structure and scope. Chapters 2–6 discuss the use of mobile network data in different steps of the four-step models, while Chap. 7 describes a framework that benefits the activity-based model and Chap. 8 presents an application of mobile network data-based travel behavior analysis in the context of tourists. Chapter 2 discusses a framework for inferring trip generation (step 1 of the four-step models) that reflects a travel demand for public transit in Senegal. As a result, a set of strategic locations for public transportation service is suggested. Chapter 3 describes a mobile network data-based framework for trip distribution modeling (step 2 of the four-step model) and evaluation of the influence of intrazonal trips in model estimation through a case study of mobile phone users in Senegal. Chapter 4 presents a method for inferring migration flows based on mobile phone users in Portugal and examines different trip distribution models (step 2 of the four-step model). Chapter 5 discusses a framework for inferring transport modes (step 3 of the four-step model) in the context of how social network influences the choice of transport mode for commuting, which is based on a case study of mobile phone users in Portugal. Chapter 6 presents and examines a set of different inference models for route assignment (step 4 of the four-step model) based on mobile network users in Lisbon, Portugal. From the activity-based approach's point of view, Chap. 7 describes how a daily activity pattern is inferred and investigates on the influence of weather conditions on the choice of daily activities, through a case study of a mobile network in Japan. Chapter 8 presents a framework for inferring touristic trips based on mobile phone users in Japan and discusses how the inferred trips are used in the context of a tourist behavior analysis, which includes a statistical analysis, such as tourist flows, top tourist destinations and origins, top destination types, top modes of transportation in terms of time spent and distance traveled, as well as a visual analytics tool developed based on this analysis. Finally, Chapter 9 provides an outlook on the future directions of the mobile network data-based urban informatics from the perspective of travel behavior research, including mobile network data characteristics, data collection, data uncertainty and privacy, and travel behavior pattern mining.

This book aims at advanced undergraduates, graduate students, researchers, and practitioners in urban informatics, particularly those who are interested in utilization of mobile network data in travel behavior research. I hope that the readers find this book useful as an overview of urban informatics and a practical guideline for using mobile network data in travel behavior research.

Chiang Mai, Thailand   Santi Phithakkitnukoon

Acknowledgments

Research studies discussed in this book are from fruitful collaborations with my colleagues over the years. So, I would like to thank Dr. Merkebe Demissie, Dr. Francisco Antunes, Dr. Rui Gomes, Dr. Carlos Bento, Dr. Zbigniew Smoreda, Dr. Tuck Leong, Dr. Teerayut Horanont, Dr. Apichon Witayangkurn, Dr. Juggapong Natwichai, Prof. Lina Kattan, Prof. Carlo Ratti, Prof. Yoshihide Sekimoto, and Prof. Ryosuke Shibasaki for their support. I would also like to thank my students, Titipat Sukhvibul, Soranan Hankaew, and Pitchaya Sakamanee, who worked with me in studies discussed in this book.

I would like to thank my colleagues at the Department of Computer Engineering, Chiang Mai University for their constant support of my teaching and research. Special thanks to Dr. Sanpawat Kantabutra (Bobby) for giving me his book *Problem Solving in Algorithms: A Research Approach*, which greatly inspired me to write my own book. This book's structure is very much inspired by his book.

I would like to express my gratitude to Prof. Carlo Ratti for introducing me to this interesting field of urban informatics research when I was a postdoctoral researcher in his Senseable City Lab at MIT, and for his long-term support of my research and career.

To my dearest wife, Raktida—thank you for all your support and encouragement. You have been my main support since day one of this book-writing journey. I'm so blessed of having you. Dear mom and dad, I wish you were still here to see this book. I miss you both so much.

Many thanks to Springer's editorial service team, Nick Zhu, Jayesh Kalleri, Ellen Seo, and Periyanayagam Leoselvakumar for helping me throughout the whole publishing process.

Contents

About the Author

Santi Phithakkitnukoon is originally from Chiang Mai, Thailand. Santi is currently an Associate Professor with the Department of Computer Engineering, Faculty of Engineering, Chiang Mai University. He received B.S. and M.S. degrees in electrical engineering from the Southern Methodist University, USA, in 2003 and 2005, respectively, and a Ph.D. in computer science and engineering from the University of North Texas, USA. Before joining Chiang Mai University, he was a Lecturer in Computing with The Open University, UK, a Research Associate with the Culture Lab (now known as Open Lab), Newcastle University, UK, and a Postdoctoral Fellow with the SENSEable City Lab, Massachusetts Institute of Technology, USA. His research interest is in urban informatics, particularly in analyzing large-scale digital footprints such as mobile phone CDRs, GPS traces, social media data, and opportunistic sensing data sources to better understand human behavior and urban dynamics.

The Overview of Mobile Network Data-Driven Urban Informatics

1

Abstract

This introductory chapter presents the key concepts of urban informatics, discusses the use of mobile network data in urban informatics, and provides an outline of the book's structure and scope. It discusses traditional methods in understanding travel behavior i.e., survey-based approach. A surveyed information is used as an input for travel demand models that predict future travel behavior and transportation needs from the current travel behavior data. A survey-based approach is however costly and time-consuming, and hence its information can be dated and limited to a particular region. Mobile network data thus emerges as a promising alternative data source that is massive in both cross-sectional and longitudinal perspectives, which provides a wider geographic coverage of travelers and a longer time of travel behavior observation. From the point of view of urban informatics, there are new opportunities as well as challenges in utilizing such data in travel behavior research. This chapter discusses the state of the art and outlines the rest of the book, which is a series of case studies done by our research teams based on mobile phone data collected from different countries to tackle different steps and approaches in travel demand modeling.

Keywords

Urban informatics · Mobile network data utilization · Travel behavior research · Travel demand modeling

© The Author(s), under exclusive license to Springer Nature Singapore Pte Ltd. 2023
S. Phithakkitnukoon, *Urban Informatics Using Mobile Network Data*,
https://doi.org/10.1007/978-981-19-6714-6_1

1.1   Urban Informatics

The term 'urban informatics' was first mentioned back in 1987 by Mark E. Hepworth in his article "The information city" [1] on page 261, as he discussed the notion of 'informatics planning' with visionary thoughts about how information and communication technologies would bring about major changes and the impact on cities. Later, in 2011, after the advent of ubiquitous computing, Forth et al. [2] described urban informatics as an interdisciplinary field of research and practice that draws on the three broad domains: people, place, and technology. People refer to all city residents that include individuals as well as groups, such as organizations and businesses. Place refers to urban sites, locales, and regions. Technology refers to information and communication technologies as well as ubiquitous computing technologies such as mobile phones, wearable devices, sensors, and internet of things (IoT) devices. Two key aspects were emphasized: (1) the new possibilities created by real-time data and ubiquitous computing, and (2) the convergence of physical and digital layers of the city. Released in 2021, the book "Urban Informatics" [3] by Shi et al. defines urban informatics as an interdisciplinary approach to understanding, managing, and designing the city using systematic theories and methods based on new information technologies, and grounded in contemporary developments of computers and communications. Three main integrated components are urban science, geomatics, and informatics. Urban science provides studies of activities, places, and flows, while geomatics provides the science and technologies for measuring dynamic spatiotemporal urban objects and managing the measured data, and informatics provides the science and statistics that facilitate the development of applications to cities.

With the unprecedented amount of user-generated data from our cities that are becoming increasingly more instrumented, the urban informatics in our own definition is an interdisciplinary and data-driven approach to understanding how city functions in various contexts from which the insights gained are used to drive the development of applications and policies for more livable and sustainable cities. It integrates urban sensing, urban analytics, and urban computing in a sustainable 'smart city' ecosystem, where the urban sensing provides mechanisms for sensing urban data, then the urban analytics provides science and statistics such as data mining, machine learning, and artificial intelligence methods to reason the data, and finally the urban computing provides computing technologies to support development of urban data-driven applications to the cities. The response from the cities is then collected by the urban sensing mechanisms, and so the smart city ecosystem continues as the urban informatics serves as its engine (see Fig. 1.1).

Urban informatics has emerged as an exciting and promising field requiring many interdisciplinary perspectives that involve computation as a central part by which methods and models are developed to generate a deeper understanding of city's functioning. The field of urban informatics is developing rapidly in the light of ubiquitous computing and sensor technologies. Sensors have scaled down their size and prices, which make them increasingly more accessible and affordable for collecting data at a large (or city) scale. This has led to big data that allows us to

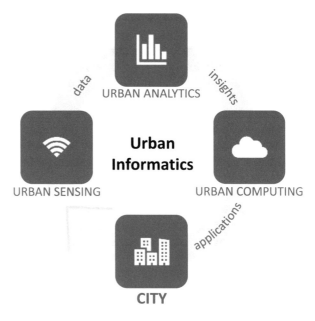

Fig. 1.1 The smart city ecosystem is driven by urban informatics, which is an integration of urban sensing, urban analytics, and urban computing

examine and validate traditional and longstanding urban system models with real-world data as well as develop new models and methods required for processing and analyzing this new large-scale sensory data.

The fact that we are now able to process these sensory data in real time has triggered the vision of real-time city [4], where city makes use of its citizen-generated data to enable real-time analysis of city life to improve its operation and management as well as to help address the problems that afflict the modern city. Real-time movement data from sensors can be exploited to monitor and better understand transportation, traffic congestion, housing, socio-economic status, air pollution, urban heating, and crime, just name a few. The movement of people defines the pace and essentially the 'rhythm' of the city. Due to its various potential benefits such as proactive urban planning, intelligent transportation systems, and traffic management, mobility studies have become central to urban informatics. A large part of urban informatics involves development of data mining and machine learning algorithms for extracting and characterizing mobility patterns from heterogenous urban data sources. Data can be gathered from sensors deployed to purposely collect such information e.g., surveillance camera for traffic count, GPS device for location tracking—i.e., active sensing paradigm. On the other hand, there is a passive sensing approach, also known as 'opportunistic sensing,' where data is intentionally collected for one purpose but can be exploited to create an opportunity that is beneficial to another purpose. A great example is mobile network data that is collected by telecom operator for billing purposes, however when exploited analytically the recorded communication logs can help reveal useful complex social network properties, while mobile users' location information can also help enhance our understanding of transport network and people's travel behavior. The opportunistic sensing data has the edge over its counterpart with its low (or no) cost as it's already been collected for something else unlike the active sensing data that comes with a price tag. The opportunistic sensing data sources thus become valuable to urban informatics and travel behavior studies, especially mobile network data as it provides a longitudinal real-life behavior of a great amount of people.

This book therefore discusses the role of mobile network data in urban informatics, particularly how the mobile network data is utilized in the mobility aspects in which approaches, models, and systems are developed for understanding travel behavior. The objectives of this book are thus to evaluate the extent to which mobile network data reflects travel behavior and to develop guidelines on how to best use such data to understand and model travel behavior. To achieve these objectives, this book attempts to evaluate the strengths and weaknesses of this data source for urban informatics and its applicability to the development and application of travel behavior models through a series of our research studies discussed in Chaps. 2–8. The following sections will discuss the current practice in travel behavior study, the use of mobile phones as sensors for collecting travel behavior data, and the use of mobile network data to infer about travel behavior.

1.2   Traditional Methods in Travel Behavior Understanding

Traditionally, household travel surveys have been used to collect travel behavior data from a sample group of households in a region. Travel demand is then quantified based on these travel surveys that are typically combined with some regionwide estimates of population and employment. Data from additional surveys such as those that focus on different transport modes, subpopulations of users, and geographies of interest are often used to provide supplemental information. Other data collection methods such as traffic counts, transit ridership estimates, and roadside interviews are also used to provide a snapshot of travel demand by mode.

Information about household members and their daily travel is collected by the household travel surveys, which is then used to determine the explanatory variables used in travel demand modeling. Typically, information collected includes when and where trips are made, the activities associated with those trips, the number of trips, trip purpose, transport modes used, travel distance and duration, and travel costs. Household and individual socioeconomic information are typically included, such as household income, household size, household life cycle, automobile accessibility, individual employment, and work locations.

Each member of the recruited household is typically asked to complete a diary of the activities and travel on a given travel day. Some recent efforts have expanded the surveys to recognize variability within a week or the number of days for which a diary needs to be completed. As it is a highly laborious and costly effort, it's not realistic to ask every household to complete the surveys. The cost of household surveys ranges between $150 and $300 per completed household, depending upon administration and technology used. So, the surveyed household travels are weighted, i.e., expanded statistically, to represent all travel behavior of the whole region, which is normally based on about 1% sampling rate. The required number of completed surveys is typically determined as the lowest number that can produce an adequate sample size that is statistically significant enough to support behavioral choice models. Survey data collection often encounter low response, nonresponse, and differential response rates. There are also some biases inherent in the sample selection, survey questionnaire, and data collection process. As a result, certain types of trips are not fully captured by the survey. Those underreported trips may include short-distance travel, nonmotorized trips, and trips made by nonrecruited households that are clustered in specific areas of a region.

Conducting a travel behavior survey roughly every 10 years is considered good practice, although many regions in different counties are unable to afford it and so do it infrequently. Each survey helps capture changes in socioeconomics, development, and travel patterns as well as improve and update the underlying travel demand model. Data from survey diaries provides records of trips and activities associated with locations at specific times of day, while surveyed data from regional sample is expanded to represent the overall regional travel behavior for analyzing and developing travel models, which combine methods of statistical sampling in local [5, 6] and national household travel surveys [7, 8] to process and infer travel at different levels of detail, such as city, region, state, and nation.

As travel demand models increasingly become more sophisticated, more detailed and higher quality input data are required. This includes not only detailed travel behavior information from household surveys, but also detailed supplemental data on transportation networks highway capacities, transit service level, land use, population, and employment.

The detailed representation of trips and activities at the household and individual levels is among the strengths of the traditional survey approach, which also include socioeconomic characteristics of the travelers, trip ends, trip purposes, transport modes used, and travel time. However, its main drawbacks are the high cost, low sampling rate, biases, inaccuracy, and outdated information.

The surveyed travel behavior information is used as an input for travel demand models that make use of current travel behavior data to predict future travel behavior and transportation needs. The two most conventional travel demand models that have played an essential role in managing and planning for transportation systems [9–11] are traditional four-step models [12] and activity-based models [13].

The four-step models have been widely accepted and traditional approach for determining transportation forecasts that include four sequential steps or models of trip generation, trip distribution, mode choice, and trip assignment. Each of the four-step models is based on household travel surveys to generate total travel and travel patterns by purpose. Trip generation model is the first step, which describes the amount of origins and destinations of trips generated in each traffic analysis zone (TAZ) by trip purpose, as a function of land use, household demographics, and other socioeconomic factors. As the second step, the trip distribution model matches the origins with destinations, which subsequently yields an origin destination table or origin-destination (O-D) matrix that describes how trips are distributed across TAZs.

In the third step, the mode choice model determines the proportion of trips between each origin and destination that use a particular transportation mode. Trip assignment is the last step, which allocates trips between origins and destinations by particular modes to routes in transportation networks, so that facility needs, costs, and benefits can be determined from the number of travelers on each route of the network. Different levels of disaggregation in estimation and application are typically used to address differences in travel behavior by time of the day, day of the week, and transportation segment.

On the other hand, the activity-based models are based on the principal that travel demand is derived from activities that people need or wish to perform, and hence the models explicitly consider travel as a derived demand in pursuit of activities, which consequently predict for individuals where and when specific activities are conducted, such as work, leisure, and shopping. Travel decisions are then formed as a component of an activity scheduling decision. The models also rely on household travel surveys and detailed journey information to extract sequences of activities and then derive daily activity patterns. Activity episode generation and scheduling processes are modeled based on an analysis of the surveyed travel behavior data [14]. As the activity-based models consider more detailed travel diary information (e.g., time and duration of trips and activities, sequences, etc.), activity-based modeling has been viewed as a more advanced approach compared to the traditional four-step modeling procedure. Although theoretically the activity-based modeling approach should provide more accurate results, the traditional four-step travel demand modeling still remains the most popular modeling approach.

Activity-based modeling approach has the advantage over the four-step modeling approach with its higher fidelity and better policy sensitivity, but the activity-based approach can be less capable of reproducing observed traffic counts when compared to the traditional four-step approach [15]. Both approaches are therefore often used to complement each other in travel demand modeling as well as provide alternatives for decision-makers in choosing an appropriate modeling approach for their transportation planning tasks.

1.3 Mobile Network-Based Travel Behavior Data Sensing

Cities are growing at unprecedented rates [16]. With the advancement in information technologies, cities have been empowered technologically, as the core systems on which they are based become instrumented and interconnected, and thus are enabling new levels of intelligence through sources of urban sensing data. High penetration of mobile network services has transformed cities into repositories of massive digital traces of human behavior with fine-grained spatial and temporal information. Mobile phones are arguably an indispensable part of modern life. People carry and interact with their mobile phones nearly all the time. As cellular network connectivity for mobile services has been recorded by telecom operators for belling, mobile phones can be utilized as personal sensors that capture individual behavior, which collectively can also sense group or regional behaviors. A wealth of data has been generated by the pervasive use of mobile phones, which can be analyzed to reveal travel behavior, patterns, and flows as it contains traces of individual user locations. Such data offers new possibilities for transportation engineers and urban planners to examine and model travel behavior [17].

With advancing ubiquitous computing technology, mobile devices (including mobile phones, tablets, and wearables) increasingly become effective active as well as passive sensors of individuals' daily mobility and activities [18]. Mobile phone penetration rate is high and increasing worldwide. There are over 5 billion unique mobile phone users in the world, while unique mobile users are growing at a rate of nearly 2% per year in 2022. A survey by Deloitte [19] confirms that mobile phone has become increasingly pervasive and indispensable with consumers worldwide are enthusiastically embracing its potential. Mobile phone penetration rates of over 90% are observed for both developed and developing countries. People increasingly interact with their phone. More than one-third of mobile users worldwide check their phone within 5 min of waking up in the morning, and 20% of them check their phone more than 50 times a day. The reliance on mobile phones seems likely to increase as more features become available.

Call detail records or CDR are logs of mobile network connectivity of an individual user, which is automatically recorded and collected by telecom operator for billing purposes. Each record contains time-stamped coordinates of the user whenever a connection to the mobile network is established, such as making or receiving a call, sending or receiving an SMS (short message service), and using the internet. In the case of voice call, caller and callee identifications as well as call duration are also recorded. Similarly, sender and receiver identifications are recorded when SMS is the service used. Not only that it can be used for billing as it is originally intended, but the collected location and time data opportunistically provides rich spatial and temporal information about human mobility patterns and travel behavior. Compared with the traditional travel surveys, these data can be gathered more often and at a much larger scale. Mobile network data is massive in both cross-sectional and longitudinal perspectives, which provides a wider geographic coverage of travelers and a longer time of travel behavior observation.

Mobile network data has been used in recent studies of human mobility, which have found that individuals are generally predictable [20, 21], unique [22], infrequently exploring new places [17], and influenced by social ties [23]. A number of useful statistical properties and models that help better understand human mobility and travel behavior still remain to be discovered from the utilization of mobile network data. Additional information gathered from GIS (geographic information system) and socioeconomic data sources, such as land use, census, road networks, employment, and POI (point of interest) that describe cities and their citizens can be advantageous for the analysis and modeling.

Travel behavior data collected by the traditional survey-based methods is becoming more costly due to its laborious procedure. As such, most countries collect such data once every 10 years or more apart. The inherent large gap between each survey can easily turn it into a dated information. Some surveyed data can also be inaccurate as they are based on individual recall of past journey details. Biases can also be caused by the sample selection, survey questionnaire, and data collection process. On the other hand, mobile phones that are carried and interacted by their users (i.e., travelers) on daily bases, passively collect a rich source of spatial and temporal information that can be translated to mobility data that describes travel behavior on a large scale.

Individual traces of mobile network connectivity that are passively collected from daily use of voice calls, text messages, and the internet, are analyzed to reconstruct the person's mobility with associated social networking information, which can then provide some of the typical outputs of regional travel demand model. Although such data can provide more accurate spatial and temporal information of individual travel on a much larger scale in both cross-sectional and longitudinal perspectives than the traditional survey-based methods, the interpretation of this new stream of data still requires development of new analysis tools and methods for efficient inference of trip count, trip purpose, activity engaged, exact location of trip's origin, destination visited, transport mode used, and route choice, just to name a few. Due to privacy concern and regulations, telecom operators are prevented from providing identifying customer information, and thereby several challenges are created from the point of view of traditional analysis. The nature of mobile phone usage and telecom operators' data collection policies also add more to these challenges. As connectivity is monitored and collected only when mobile service is used i.e., the user connects to the network for a service, there are discontinuities in user mobility causing unreported trips, which can be difficult to identify. There can be scenarios where a traveler is carrying multiple mobile phones or a single phone is used by multiple travelers, in which it can also be difficult for the analyst to identify. Market share may play a role in the analysis if the mobile network data isn't obtained from all telecom operators. In any case, it is rare to obtain data from all operators, unless the analysis is supported by government or telecommunication authority who can administer such data. The purpose of each trip needs to be inferred, for which algorithms and methods for extracting individual trips, exact location of trip origin and destination, as well as intermediate stops need to be developed. Other challenges can also emerge due to the difficulty of its specific inference problem, such as transport modes used and route choices taken by the travelers. Translating mobile phone location traces into route choices isn't trivial, which may involve applying map matching algorithms [24] and GIS techniques [25] with additional transportation network data. Likewise, heuristic algorithms for identifying transport modes need to be developed, which can be a subsequential step from the route choice inference by taking into account the inferred routes to figure out possible modes of transportation used. The inferred intermediate stops are then used for further analysis and inference of activities engaged and daily activity patterns by using additional information, such as land use, POIs, and social events for instance.

Looking on the bright side, these challenges also have open up unique opportunities for research across all domains of the urban informatics. Although the mobile network data-derived travel behavior information isn't as descriptive as survey-based data that provides specific trip details, its longitudinality allows the analyst to infer travel behavior reasonably with statistical significance. Long period data allows an observation of multiple repeated trips that may be used to recognize commuting and non-commuting journeys from trip patterns, and hence it may lead to constructing less invasive travel diaries over an extended period. With additional geospatial data, users who travel regularly to specific types of places for work or else, e.g., hospital, university, shopping, and so on, may be recognized by their personas along with their travel behavior, from which the purpose of the trips and likely activities engaged may be inferred intuitively. This mobile data-driven approach that offers a longer-term observation may yield a better travel and activity data set for at least some aspects of daily travel when compared with diary data collected for a small sample during a limited period.

1.4   Mobile Network Data-Based Travel Behavior Inference

Since the traditional approach to travel demand modeling is based on a small survey sample of daily travel, it relies on developing analytical procedures and making inferences that allow the use of a small sample to represent the daily travel in a region. On the other hand, the mobile network data although provides a wider geographic coverage of travelers and a longer time of travel behavior observation, it doesn't record trips. New procedures, methods, and algorithms for inferring trips and activities is thus needed. Trip ends as well as the basis of spatial and temporal patterns can be inferred from the sequence of mobile phone connectivity data.

The use of mobile phone network data has shown its benefit in various contexts, such as human mobility, epidemiology, and social network analysis. In human mobility, mobile network data has been used to show that people's traveling is regular and highly predictable [26], and most people travel only short distances as well as have a high tendency of revisiting the same places [20]. Epidemiology is benefited from understanding human mobility by analyzing mobile network data, such as an understanding of how human travel patterns contributes to Malaria spread [27] and how human mobility can be used to model fundamental spreading patterns that characterize a mobile virus outbreak [28]. In the context of social network analysis, the mobile network data has been explored in modeling social network structure, such as a scaling ratio in social group sizes [29] and a change in social network structure over time due to a migrations [30] or behavior adaptation [31], as well as a discovery of a link between social variety and economic development [32]. Moreover, the mobile network data has also been explored in the contexts of incident and traffic management [33], privacy preservation [34], and socio-geography of human mobility [35].

In order to use a mobile network data to make inferences about travel behavior such as trips and activities, a set of procedures need to be considered. To infer a trip, a series of methods must be developed to firstly extract a trajectory from mobile network connectivity by making use of both timestamp and connected location records, i.e., a temporal sequence of connected locations. Secondly, another method needs to be developed to transform an extracted trajectory into trips, which are subsequences of the trajectory separated by stop locations. Such rule may involve setting strategic spatial and temporal thresholds for identifying a stop location—i.e., a location where a traveler makes a stop and stays at the location over a period. This may be an attempt to detect a no change or small changes in connected locations within a radius distance over some period [36, 37]. The detected stop locations can be treated as an origin of the next trip or a destination of the previous trip. These origin and destination pairs can also be utilized for the trip generation and trip distribution modeling [38, 39], which are the first and second steps of the four-step models. Origins and destinations can be clustered into a traditional fixed TAZ, such as the official boundaries [40], grids [41] and Voronoi-based zones [42], or an adaptive zoning scheme where O-D areas vary with spatial demand density [43].

Trips can be categorized into two main types: commuting and non-commuting. Commuting is a common and recurring travel between one's residence and work-place, which accounts for most of the trips made by individuals and is one of the key causes of traffic congestion. Identifying a commuting trip can be done by recognizing residence and work locations, which may be locations with a high connectivity during usual hours that most people spend at home and workplace, i.e., late night and business hours, respectively [23, 44]. Non-commuting trip, on the other hand, is an infrequent and non-routine travel, e.g., migration, summer vacation, weekend getaway, and business trip. Migration is a travel for relocation of residence, which can be permanent or temporary. The key is to detect the change of residence and determine if it is a migration. Inference models for residential location and its change, as well as a migration definition are needed. The type of migration may be determined by a rule-based classification. For example, a permanent migration may be defined as it occurs when there is a change of residence from one location to another and no return, while a temporary migration takes place in scenario where there is a return to former residential location [45, 46]. For touristic trips, it can be further categorized into short and long-term trips, such as a 1-day trip and weekend getaway, respectively. Its destination location, time spent before its return to the residential location are among the key indicators for determining a touristic trip's type.

In order to perform a route assignment i.e., the fourth step of the four-step models, another inference method needs to be developed to map the detected trips onto a road network, which may involve applying a map matching algorithm [47, 48] or utilizing a route planning service API, e.g., matching an inferred trip (a location sequence) to the most probable route suggested by a mapping service such as Google Maps [49–51]. Although it is the last step in the four-step models, the route assignment typically appears in this stage from the travel behavior inference modeling's point of view as a route information necessitates making an inference about a transport mode, which is the third step in the four-step models. Developing a transport mode inference model may consider a travel speed calculation [52], utilizing a mapping service API [53], semi-supervised learning approach such as clustering and Bayesian inference [54], or rule-based approach [55].

From the activity-based modeling's point of view, a method needs to be developed for inferring an activity engaged by the traveler at a stop location. This may involve additional geospatial data such as land use, building footprints, and POIs, which help characterize an area in which an information about likely activities can be gathered. The most probable activity engaged by a traveler who visits the area may be identified based on the activity information gathered from the area [56]. Daily activity pattern or activity-based mobility may then be extracted based on individual visit frequencies to different locations.

These inferred trips and activities can then be used to analyze various aspects of travel behavior, which may involve different statistical measurements, such as information entropy, correlation coefficient, clustering coefficient, and so on. Nonetheless, the aforementioned procedures and methods may encounter problems under different scenarios. For example, a mobile device is not used during traveling, so there's no connectivity data recorded, and hence no information about such trips and activities can be inferred. Incorrect inference can be made if the mobile device is used by a someone whose life schedule is unusual, such as digital nomad, retiree, unemployed person, telecommuter, or grave-yard shift worker. A device that is used by multiple users can also cause incorrect inferences. However, these issues may not be significant if the expanded results are comparable to the expended trips from household surveys or from regional models at an aggregate level. So, it is highly recommended to validate preliminary results, if possible, prior to the development of subsequential inference models. It may be a comparison test against census information, such as population density and regional travel survey for instance. Ultimately, the use of mobile network data for inferring travel behavior, such as trips and activities requires a sufficient amount of connectivity logs (e.g., voice calls, SMS, internet, signaling) to be recorded for each mobile device over a period of time. The number of mobile devices, spatial coverage, and observation period are among the most indicative characteristics of the data's representativeness for the actual regional travel behavior.

As this book aims to provide researchers and practitioners with an understanding of urban informatics and an important role that the mobile network data plays in advancing its state of the art from the perspectives of travel behavior research, the following chapters discuss inference models and frameworks, which our research teams have developed over the years for trips and activities that characterize travel behavior through real-world case studies. The next seven chapters address how a mobile network data can be used in the four-step models, the activity-based approach, and travel behavior analysis, which serve as research baselines in mobile network data-driven urban informatics. Chapters 2–6 discuss a use of mobile network data in different steps of the four-step models. To begin with, Chap. 2 discusses a framework for inferring trip generation that reflects a travel demand for public transit in Senegal. As a result, a set of strategic locations for public transportation service is suggested. Chapter 3 describes a mobile network data-based framework for trip distribution modeling and evaluation of the influence of intrazonal trips in model estimation through a case study of mobile phone users in Senegal. Chapter 4 presents a method for inferring migration flows based on mobile phone users in Portugal and examines different trip distribution models. Chapter 5 discusses a framework for inferring transport modes in the context of how social network influences the choice of transport mode for commuting, which is based on a case study of mobile phone users in Portugal. Chapter 6 presents and examines a set of different inference models for route assignment based on mobile network users in Lisbon, Portugal. From the activity-based approach's point of view, Chapter 7 describes how a daily activity pattern is inferred and investigates on the influence of weather conditions on the choice of daily activities, through a case study of a mobile network in Japan. Chapter 8 presents a framework for inferring touristic trips based on mobile phone users in Japan and discusses how the inferred trips are used in the context of a tourist behavior analysis, which includes a statistical analysis, such as tourist flows, top tourist destinations and origins, top destination types, top modes of transportation in terms of time spent and distance traveled, as well as a visual analytics tool developed based on this analysis. Finally, Chap. 9 provides an outlook on the future directions of the mobile network data-based urban informatics from the perspective of travel behavior research, including mobile network data characteristics, data collection, data uncertainty and privacy, and travel behavior pattern mining.

References

1. Hepworth ME. The information city. Cities. 1987;4(3):253–62.
2. Foth M, Choi JHJ, Satchell C. Urban informatics. In: Proceedings of the ACM conference on computer supported cooperative work (CSCW'11); 2011. p. 1–8. https://doi.org/10.1145/1958824.1958826.
3. Shi W, Goodchild M, Batty M, Kwan M-P, Zhang A. Urban informatics. Singapore: Springer; 2021.
4. Kitchin R. The real-time city? Big data and smart urbanism. GeoJournal. 2014;79:1–14. https://doi.org/10.1007/s10708-013-9516-8.
5. Daganzo CF. Optimal sampling strategies for statistical models with discrete dependent variables. Transp Sci. 1980;14(4):324–45. https://doi.org/10.1287/trsc.14.4.324.
6. Smith ME. Design of small-sample home-interview travel surveys. Transp Res Board. 1979;701:29–35.
7. Stopher PR, Greaves SP. Household travel surveys: where are we going? Transp Res Part A Policy Pract. 2007;21(5):367–81. https://doi.org/10.1016/j.tra.2006.09.005.
8. Richardson AJ, Ampt ES, Meyburg AH. Survey methods for transport planning. Parkville, VIC: Eucalyptus Press; 1995.
9. Manheim ML. Fundamentals of transportation systems analysis. Cambridge, MA: MIT Press; 1979.
10. Ben-Akiva M, Lerman S. Discrete choice analysis. Boston, MA: MIT Press; 1985.
11. Ortúzar J, Willumsen LG. Modelling transport. Chichester, UK: Wiley-Blackwell; 2011.
12. Mcnally MG. The four step model. Handb Transp Model. 2007. https://escholarship.org/uc/item/0r75311t.
13. Castiglione J, Bradley M, Gliebe J. Activity-based travel demand models: a primer. Washington, DC: The National Academies Press; 2014.
14. Bhat CR, Koppelman FS. Activity-based modeling of travel demand. In: Hall RW, editor. The handbook of transportation science. Norwell, MA: Kluwer Academic; 1999. p. 35–61.
15. Zhong M, Shan R, Du D, Lu C. A comparative analysis of traditional four-step and activity-based travel demand modeling: a case study of Tampa, Florida. Transp Plan Technol. 2015;38 (5):517–33. https://doi.org/10.1080/03081060.2015.1039232.
16. United Nations. World urbanization prospects. New York; 2014. https://doi.org/10.4054/DemRes.2005.12.9.

17. Jiang S, Fiore GA, Yang Y, Ferreira J, Frazzoli E, González MC. A review of urban computing for mobile phone traces: current methods, challenges and opportunities. In: Proceedings of the ACM SIGKDD international conference on knowledge discovery and data mining; 2013. p. 1–9. https://doi.org/10.1145/2505821.2505828.

18. Blondel VD, Decuyper A, Krings G. A survey of results on mobile phone datasets analysis. EPJ Data Science. 2015. https://doi.org/10.1140/epjds/s13688-015-0046-0.

19. Wigginton C, Curran M, Brodeur C. Deloitte: global mobile consumer trends. 2017. [Online]. https://www2.deloitte.com/content/dam/Deloitte/us/Documents/technology-media-telecommunications/us-global-mobile-consumer-survey-second-edition.pdf.

20. González MC, Hidalgo CA, Barabási AL. Understanding individual human mobility patterns. Nature. 2008;453:779–82. https://doi.org/10.1038/nature06958.

21. Song C, Koren T, Wang P, Barabási AL. Modelling the scaling properties of human mobility. Nat Phys. 2010;6:818–23. https://doi.org/10.1038/nphys1760.

22. Calabrese F, Diao M, Di Lorenzo G, Ferreira J, Ratti C. Understanding individual mobility patterns from urban sensing data: a mobile phone trace example. Transp Res Part C Emerg Technol. 2013;26:301–13. https://doi.org/10.1016/j.trc.2012.09.009.

23. Phithakkitnukoon S, Smoreda Z, Olivier P. Socio-geography of human mobility: a study using longitudinal mobile phone data. PLoS One. 2012;7(6):e39253. https://doi.org/10.1371/journal.pone.0039253.

24. Chao P, Xu Y, Hua W, Zhou X. A survey on map-matching algorithms. In: Borovica-Gajic R, Qi J, Wang W, editors. Databases theory and applications. Lecture notes in computer science, vol. 12008; 2020. p. 121–33. https://doi.org/10.1007/978-3-030-39469-1_10.

25. Hochmair HH. Introducing geographic information systems with ArcGIS: a workbook approach to learning GIS. 3rd ed. Hoboken, NJ: Wiley; 2013.

26. Song C, Qu Z, Blumm N, Barabási AL. Limits of predictability in human mobility. Science. 2010;327(5968):1018–21. https://doi.org/10.1126/science.1177170.

27. Wesolowski A, et al. Quantifying the impact of human mobility on malaria. Science. 2012;338 (6104):267–70. https://doi.org/10.1126/science.1223467.

28. Wang P, González MC, Hidalgo CA, Barabási A-L. Understanding the spreading patterns of mobile phone viruses. Science. 2009;324(5930):1071–6. https://doi.org/10.1126/science.1167053.

29. Phithakkitnukoon S, Dantu R. Mobile social group sizes and scaling ratio. AI Soc. 2011. https://doi.org/10.1007/s00146-009-0230-5.

30. Phithakkitnukoon S, Calabrese F, Smoreda Z, Ratti C. Out of sight out of mind – how our mobile social network changes during migration. In: 2011 IEEE third international conference on privacy, security, risk and trust and 2011 IEEE third international conference on social computing; 2011. p. 515–20. https://doi.org/10.1109/PASSAT/SocialCom.2011.11.

31. Eagle N, de Montjoye Y-A, Bettencourt LMA. Community computing: comparisons between rural and urban societies using mobile phone data. In: 2009 international conference on computational science and engineering; 2009. p. 144–50. https://doi.org/10.1109/CSE.2009.91.

32. Eagle N, Macy M, Claxton R. Network diversity and economic development. Science. 2010;328(5981):1029–31. https://doi.org/10.1126/science.1186605.

33. Steenbruggen J, Borzacchiello MT, Nijkamp P, Scholten H. Mobile phone data from GSM networks for traffic parameter and urban spatial pattern assessment: a review of applications and opportunities. GeoJournal. 2013;78(2):223–43. https://doi.org/10.1007/s10708-011-9413-y.

34. Yang J, Dash M, Teo SG. PPTPF: privacy-preserving trajectory publication framework for CDR mobile trajectories. ISPRS Int J Geo-Inf. 2021;10(4):1–19. https://doi.org/10.3390/ijgi10040224.

35. Scott DM, Dam I, Páez A, Wilton RD. Investigating the effects of social influence on the choice to telework. Environ Plan A Econ Sp. 2012;44(5):1016–31. https://doi.org/10.1068/a43223.

36. Phithakkitnukoon S, Horanont T, Witayangkurn A, Siri R, Sekimoto Y, Shibasaki R. Understanding tourist behavior using large-scale mobile sensing approach: a case study of

mobile phone users in Japan. Pervasive Mob Comput. 2015;18. https://doi.org/10.1016/j.pmcj.
2014.07.003.

37. Luo T, Zheng X, Xu G, Fu K, Ren W. An improved DBSCAN algorithm to detect stops in
individual trajectories. ISPRS Int J Geo-Inf. 2017;6(3):63. https://doi.org/10.3390/ijgi6030063.

38. Demissie MG, Phithakkitnukoon S, Kattan L. Trip distribution modeling using mobile
phone data: emphasis on intra-zonal trips. IEEE Trans Intell Transp Syst. 2018;20(7):
2605–17. https://doi.org/10.1109/TITS.2018.2868468.

39. Demissie MG, Phithakkitnukoon S, Kattan L, Farhan A. Understanding human mobility
patterns in a developing country using mobile phone data. Data Sci J. 2019;18(1). https://doi.
org/10.5334/dsj-2019-001.

40. Demissie MG, Antunes F, Bento C, Phithakkitnukoon S, Sukhvibul T. Inferring origin-
destination flows using mobile phone data: a case study of Senegal. In: 2016 13th international
conference on electrical engineering/electronics, computer, telecommunications and information technology (ECTI-CON); 2016. p. 1–6. https://doi.org/10.1109/ECTICon.2016.7561328.

41. Phithakkitnukoon S, Veloso M, Bento C, Biderman A, Ratti C. Taxi-aware map: identifying
and predicting vacant taxis in the city. In: de Ruyter B, et al., editors. *Ambient intelligence*,
Lecture notes in computer science, vol. 6439; 2010. p. 86–95. https://doi.org/10.1007/978-3-
642-16917-5_9.

42. Gundlegård D, Rydergren C, Breyer N, Rajna B. Travel demand estimation and network
assignment based on cellular network data. Comput Commun. 2016;95(1):29–42. https://doi.
org/10.1016/j.comcom.2016.04.015.

43. Mungthanya W, et al. Constructing time-dependent origin-destination matrices with adaptive
zoning scheme and measuring their similarities with taxi trajectory data. IEEE Access. 2019;7.
https://doi.org/10.1109/ACCESS.2019.2922210.

44. Zagatti GA, et al. A trip to work: estimation of origin and destination of commuting patterns in
the main metropolitan regions of Haiti using CDR. Dev Eng. 2018;3:133–65. https://doi.org/10.
1016/j.deveng.2018.03.002.

45. Phithakkitnukoon S, Calabrese F, Smoreda Z, Ratti C. Out of sight out of mind – how our
mobile social network changes during migration. In: Proceedings – 2011 IEEE international
conference on privacy, security, risk and trust and IEEE international conference on social
computing, PASSAT/SocialCom 2011; 2011. p. 515–20. https://doi.org/10.1109/PASSAT/
SocialCom.2011.11.

46. Hankaew S, Phithakkitnukoon S, Demissie MG, Kattan L, Smoreda Z, Ratti C. Inferring and
modeling migration flows using mobile phone network data. IEEE Access. 2019;7(1):
164746–58. https://doi.org/10.1109/ACCESS.2019.2952911.

47. Bonnetain L, Furno A, Krug J, El Faouzi NE. Can we map-match individual cellular network
signaling trajectories in urban environments? Data-driven study. Transp Res Rec. 2019;2673
(7):74–88. https://doi.org/10.1177/0361198119847472.

48. Jagadeesh GR, Srikanthan T. Online map-matching of noisy and sparse location data with
hidden Markov and route choice models. IEEE Trans Intell Transp Syst. 2017;18(9):2423–34.
https://doi.org/10.1109/TITS.2017.2647967.

49. Jundee T, Kunyadoi C, Apavatjrut A, Phithakkitnukoon S, Smoreda Z. Inferring commuting
flows using CDR data: a case study of Lisbon, Portugal. In: UbiComp/ISWC 2018 – adjunct
proceedings of the 2018 ACM international joint conference on pervasive and ubiquitous
computing and proceedings of the 2018 ACM international symposium on wearable computers;
2018. p. 1041–50. https://doi.org/10.1145/3267305.3274159.

50. Sakamanee P, Phithakkitnukoon S, Smoreda Z, Ratti C. Methods for inferring route choice of
commuting trip from mobile phone network data. ISPRS Int J Geo-Inf. 2020;6(5):306. https://
doi.org/10.3390/ijgi9050306.

51. Wang H, Calabrese F, Di Lorenzo G, Ratti C. Transportation mode inference from anonymized
and aggregated mobile phone call detail records. In: 13th international IEEE conference on
intelligent transportation systems, proceedings (ITSC); 2010. p. 19–22. https://doi.org/10.1109/
ITSC.2010.5625188.

52. Lwin KK, Sekimoto Y. Identification of various transport modes and rail transit behaviors from mobile CDR data: a case of Yangon City. Asian Transp Stud. 2020;6(100025):1–12. https://doi.org/10.1016/j.eastsj.2020.100025.

53. Phithakkitnukoon S, Sukhvibul T, Demissie M, Smoreda Z, Natwichai J, Bento C. Inferring social influence in transport mode choice using mobile phone data. EPJ Data Sci. 2017;6(11). https://doi.org/10.1140/epjds/s13688-017-0108-6.

54. Bachir D, Khodabandelou G, Gauthier V, El Yacoubi M, Puchinger J. Inferring dynamic origin-destination flows by transport mode using mobile phone data. Transp Res Part C Emerg Technol. 2019;101:254–75. https://doi.org/10.1016/j.trc.2019.02.013.

55. Qu Y, Gong H, Wang P. Transportation mode split with mobile phone data. In: IEEE conference on intelligent transportation systems, proceedings (ITSC 2015); 2015. p. 285–9. https://doi.org/10.1109/ITSC.2015.56.

56. Phithakkitnukoon S, Horanont T, Di Lorenzo G, Shibasaki R, Ratti C. Activity-aware map: identifying human daily activity pattern using mobile phone data. In: International workshop on human behavior understanding. Berlin: Springer; 2010. p. 14–25.

Inferring Passenger Travel Demand Using Mobile Phone CDR Data

2

Abstract

Urban transportation is a key issue worldwide, especially in Sub-Saharan Africa where there has been a rapid increase in population and at the same time, a lack of infrastructure which include railways, airways and roads. People's mobility in these African nations is mostly provided by bus services and a large-scale informal public transportation system known as paratransit (for example, car rapides in Senegal, Tro Tros in Ghana, taxis in Uganda and Ethiopia, and Matatus in Kenya). This brings up the need for transport demand estimation, which is a challenging task, particularly in developing countries. The main reason for the challenge is that the estimation methods usually require large datasets which can be quite difficult, costly, and time-consuming to collect. When it comes to demand estimation, important factors include the accuracy and transparency of data. Accurate data can help us identify trends in passenger demand so that better informed decisions about future investments in infrastructure and capacity can be made. In this chapter, we discuss how passenger demand for public transportation services can be estimated using mobile phone network data. Based on the inferred travel demand, strategic locations for public transportation services such as paratransit and taxi stands can then be suggested accordingly. This chapter is inspired by our original research work done by Demissie et al. (IEEE Trans Intell Transp Syst. 2016;17(9):2466–78; 13th international conference on electrical engineering/electronics, computer, telecommunications and information technology (ECTI-CON). IEEE; 2016).

Keywords

Call detail records · Trip generation inference · Paratransit · Travel demand · Public transport · Transit route

2.1   Motivation and State of the Art

The rate of urbanization around the world is increasing steadily. In 2010, more than half of the world's population lived in cities [1]. Cities in Africa are also quickly expanding. In 2011, the population of Africa surpassed 1 billion people, and by 2050, it is expected to reach 2.4 billion [2]. Urbanization goes hand in hand with population growth. Between 1960 and 2011, Africa's urban population increased from 19 to 39%, and by 2040, half of the continent's inhabitants will live in cities [2]. This is due to the fact that cities, as opposed to rural areas, have more options for resolving social and economic issues. But as more people move to cities, additional residences, stores, schools, health centers, roads, and public transit are needed. A lot of cities have limited resources to adapt to the degree of change in urban regions. As the population of cities grows, more travel demand has resulted in issues such as traffic congestion, parking difficulties, traffic accidents, public transportation inadequacy, and environmental issues [3].

Apart from conventional buses, the majority of cities in Sub-Saharan Africa rely on large-scale informal/flexible transportation services known as paratransit (for example, car rapides in Senegal, Tro Tros in Ghana, taxis in Ethiopia, and Matatus in Kenya, among others). Paratransit is a phrase that has been used to describe "those modes of intra-urban passenger transportation that are available to the public and are separate from conventional transit (scheduled bus and rail) and can operate across the highway and transit system" [4]. Paratransit services are offered in developed nations in the form of on-demand responsive transit, dial-a-ride, and dial-a-ride transit schemes, which are commonly used by people with limited mobility. In underdeveloped nations, paratransit services are offered to the general public on a bigger scale, generally through unregulated operators in the informal sector. In other circumstances, however, this service is offered by regulated operators in the formal sector.

Paratransit services in developing nations, on the whole, have the following characteristics: they operate without a set schedule and are not limited in terms of routes or areas in which they can operate; the vehicles are typically small, ranging from 4-seat sedans to 17–35-seat midibuses; fares are usually set by the city government; and the services are provided by private operators [5, 6]. Despite the good effects of both paratransit and large-bus services, there are still unfathomable instances for enhancing public transportation quality, reliability, accessibility, and coverage in developing countries. Despite the need for such information to properly manage their urban public transportation systems, cities in developing nations have struggled to grasp existing and future motions and dynamics of their urban systems. The fundamental reason is that most developing cities lack the financial resources to collect the detailed data required for transportation planning. For example, governments in the Sub-Saharan African region have minimal data for transportation planning, most cities do not perform traffic counts on a regular basis, and no city in the sub-Saharan African region has household travel surveys except for cities in South Africa [7].

Researchers have been looking into approaches to produce large-scale urban sensing by utilizing the growing capabilities of the cellular networks system in recent years. When a mobile phone user makes a call or sends a text message, it is simple to estimate their location. Mobile phone penetration has risen at an exponential rate over the last decade. In 2013, there were almost as many mobile phone subscriptions as there were people on the planet, with 96% of active mobile phone cards per 100 inhabitants worldwide, 128% in affluent nations, and 89% in developing countries [8].

Having seen this leads us to investigate the use of mobile phone network data to dynamically infer urban mobility trends in order to improve transportation planning. To begin, we concentrate on extracting prominent home and destination anchors in order to get a sense of travel demand, i.e., the broad inward and outward flow of people and vehicles between various regions. Our second approach is to link this data to the geography of human movement as determined by a city's public bus network. This allows us to recommend key locations for public transportation services such as bus routes, paratransit stops, and taxi pick-up and drop-off locations based on actual travel demand.

Traditional transportation planning relies on an understanding of current and future issues related with urban growth, such as how much travel will be generated, where these trips will occur, and by which mode and on which routes. Estimating and forecasting demand is one of the most difficult aspects of public transportation planning. Planners must predict public transportation demand for a variety of reasons; such as for route extensions, new route implementation, new modes (e.g., Bus Rapid Transit—BRT)/services, scheduling changes, ridership forecasting as part of a short and long-term strategy, and others [9].

Data from new sources is increasingly being used to acquire a better understanding of urban public transportation networks. The city of Dublin, for example, investigated the core causes of traffic congestion in its public transportation network using traditional data and GPS bus traces. The outcome allows traffic controllers to see the present status of the entire bus network at a glance and rapidly identify places where delays are occurring that require further investigation [10]. Smart card transactions and automatic vehicle location data are also being used to study bus passenger travel behavior, such as estimating origin and destination [11], designing personalized travel information systems for individual passengers [12], and generating knowledge about future public transportation access patterns [13].

Mobile phone users, on the other hand, generate a massive amount of passive and active mobile positioning data that, when combined, can provide us with information about the presence and movement of individuals in a given region. Various research projects have focused on the use of passive mobile positioning data, which is automatically recorded in mobile operators' memory files for call activities or device movements in the network. These studies look at a variety of topics, such as travel time and speed estimation [14]; road usage patterns assessment [15]; mobile phone traffic and vehicular traffic correlation [16, 17]; traffic status detection [18]; congestion detection [19]; and route classification [20].

Another type of data from mobile phone users are active or application-based data, in which a smartphone runs dedicated software that reports its location to a server outside the cellular network. The Mobile Century project, which used GPS-equipped Nokia telephones to collect continuous position and speed profiles of automobiles [21], has undertaken a preliminary effort in this area. Other investigations by Hongsakham et al. [19] and Puntumapon and Pattara-atikom [22] used a Nokia mobile phone equipped with cellular probe software to infer traffic congestion and distinguish pedestrians from sky train passengers.

A number of studies have been conducted on the use of mobile data to estimate travel demand. Çolak et al. [23] and Alexander et al. [24] used mobile network data to create trips that were classified by time of day and purpose (e.g. home-based-work). The total trip matrices for each case study city are then generated using census data. To predict many aspects of travel demand, Toole et al. [25] combined mobile network data with census, surveys, open and crowd-sourced geospatial data, including road network performance and road categorization analyses. Alexander and Gonzlez [26] created a novel approach to assess the demand and congestion consequences of a new mode of transportation based on different adoption levels and hourly OD trips calculated from smartphone data. Jiang et al. [27] developed an activity-based model based on mobile data in order to generate overall daily activity patterns.

It is critical to evaluate travel patterns in order to determine public transportation demand and provide appropriate solutions for a community. People's travels' origins and destinations are among the most essential of these patterns. Previous research has used mobile network data to estimate origin-destination distances [28–30]. Traditionally, subjects in origin-destination surveys have been required to record data on their travel patterns, including where they are moving to and from during the observation period (typically, passengers travel for only 1 day per route or a week); how they're doing so, and why they're doing it. Mobile network data has been employed in this domain to determine land usage patterns, which is useful information for estimating the trip-generation property of various regions [31–33].

Despite this, the majority of previous research focused on passenger travel patterns between home and work. According to Ahas et al. [34], with rising individual mobility, the dominance of home and work anchors has waned, and people now spend a significant amount of time in other places. Some researchers looked into the number of sites where a person spends a lot of time and/or goes on a regular basis [35, 36]. These studies, on the other hand, are unable to link visits between users' homes and their top destination locations. Mobile network data was used in a study by Berlingerio et al. [37] to extract transport demand and use an optimization technique to identify new transit routes with the goal of reducing travel and wait times.

Our research also aims to broaden our understanding of how we might use mobile network data to extract people's travel patterns between their homes and popular destinations. However, in this research of new transit route suggestions, temporal variables (such as access, wait, and travel times) are not taken into account. Given Senegal's low public transportation coverage, we prioritize meeting the criteria for public transportation coverage and accessibility. In this chapter, we will show how mobile network data is mined to identify locations in Senegal with high transport demand but little or non-existent public transportation coverage and accessibility. We then make a suggestion on advantageous locations for public transportation services (paratransit and taxi stands, transit routes, and strategic locations for potentially constructed high-order public transportation systems, such as BRT) based on the information gathered.

2.2   Case Study Area and Dataset

Dakar (Senegal) is used as a case study to infer passenger travel demand using mobile phone network data. We will also have a look at the transit profile of Senegal. The transit profile for the case study area is based on different factors including population density and proximity to a transit stop.

2.2.1   Case Study Area

Senegal is a West African country with a total land area of 196,712 square kilometers. Senegal had an estimated population of 13,508,715 in 2013. Senegal is organized into 14 regions (Fig. 2.1), which are further broken down into 45 departments and 123 arrondissements (district). Our research will be focused on Dakar, Senegal's city, and the surrounding area. The Dakar region is organized into four departments (administrative units with no political power), each of which is divided into ten arrondissements, with a total population of 3,137,196 [38].

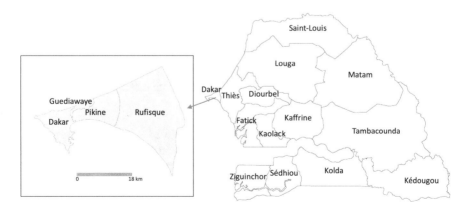

Fig. 2.1 Map of Dakar (**a**) which is a city in Senegal (**b**)

2.2.2   Transit Profile of Case Study Area

The state-owned enterprise, *Compagnie senegalaise de transformation et de conditionnement*, which was founded in the late 1940s and later renamed SOTRAC (Societ des Transports en Commun du Cap Vert) in 1971, provided transportation throughout Senegal, mainly in Dakar. Because of its restricted fleet size and governments' failure to give sufficient funding to continue its operation, SOTRAC was unable to meet the growing demand. This resulted in a decrease in service coverage and quality, leading to the company's closure in 1998 [38]. In March 1997, the Senegalese government established a new coordinating body to handle Dakar's urban transport: the Executive Council of Urban Transport in Dakar (CETUD). In Dakar, the following public transportation options are now offered.

2.2.2.1 Bus Service

Bus service is provided by Dakar Dem Dikk (DDD), a commercial public transportation firm founded in 2000 to meet Dakar's growing transportation demand. DDD began operations with a fleet of 60 buses. In 2004, DDD added 360 new buses to its fleet, allowing it to serve 17 urban routes in and around Dakar [39].

Paratransit service: The failure of large-bus services to meet the expanding mobility needs of inhabitants in Dakar prompted the development of the paratransit service, which is known by the names Ndiaga Ndiaye and Cars rapides. The typical paratransit van may transport up to 25 people (cars rapides). Midibuses with a capacity of up to 35 passengers are also available (Ndiaga Ndiaye). Paratransit in Dakar operates on a flexible route with no set schedule. There are, however, well-known large stations that can be considered beginnings and endpoints for a specific route, as well as frequently understood intermediate stations. These intermediate stations are not usually marked (there are no official station stands), but they are well-known among locals as regular pauses. It is usually possible to get on and off at any point between the point of origin and the point of destination. As a result, the current paratransit-based public transportation system lacks reliability, has a wide range of charges, and lacks route alignment schemes and timetables. Paratransit shares road space with pedestrians and street vendors, forming long lineups that frequently block junctions and disrupt traffic flow in both directions.

2.2.2.2 Taxi Service

Taxi cabs with a capacity of four persons are available 24 h a day, 7 days a week. Taxis have a wide range of prices. From midnight to 5:00 a.m., there is usually an additional fee. Because taxi meters are either missing or broken, the pricing is frequently subject to negotiation between the rider and the driver. Walking, cycling, and other non-motorized forms of transportation are also available in Dakar and its environs. There's also a train service provided by Petit Train de Banlieue that connects Dakar with larger cities in Senegal.

Dakar, being the country's main city, has long struggled with traffic congestion and a shaky public transport infrastructure. The government and municipal governments have recently paid some attention to implementing initiatives for improving urban mobility, such as developing the capacity of transport actors, renewing the public transport fleet, rehabilitating the city train, and improving critical principal highways. While significant progress has been made in alleviating severe traffic congestion, these techniques have not yielded the desired outcomes in terms of enhancing public transport systems [40].

There is clearly still room to implement further transport demand management strategies in Dakar to improve mobility. To execute these measures, a precise description of the city's mobility patterns and activities is required, but mobility surveys are scarce due to government budget constraints.

2.2.3 Dataset

2.2.3.1 Mobile Network Data

Senegal has a large population of mobile phone users. In 2012, there were 79 active mobile telephone cards for every 100 Senegalese people [41]. We also analyzed mobile communication data provided by SONATEL and Orange as part of the D4D Challenge. In 2012, SONATEL had a 61% market share in Senegal [42]. The data is based on Call Detail Records (CDRs) of SONATEL subscribers' mobile phone calls and text messages from January 7th to January 20th, 2013. This data is anonymous mobility data for 300,000 randomly picked users over the course of 2 weeks, with location recording down to the cell tower level (See Table 2.1).

The original dataset had almost 9 million unique aliased mobile phone numbers, and Orange Labs performed additional data processing to retain customers that satisfied the following two criteria [41]: (i) users who have had more than 75% of their days with interactions in a given time; and (ii) users who have had less than 1000 interactions each week on average. Machines or shared phones were assumed to represent the users who had more than 1000 interactions per week. We looked at the data from 1666 base stations that handled cellular traffic. The real geographical coordinates of the base stations were not provided by SONATEL for commercial and privacy reasons.

To make it more difficult to re-identify users, each base station's new position is allocated evenly throughout its Voronoi cell (the region consisting of all points closer to that antenna than to any other) [42]. Table 2.2 depicts a sample of new base station sites (latitude, longitude).

Table 2.1 Sample of mobile phone trajectories at the base station level	User ID	Timestamp		Base station ID
	1	18-03-2013	21:30:00	716
	1	18-03-2013	21:40:00	718
	1	19-03-2013	20:40:00	716

Table 2.2 Sample of base station position data

Base station ID	Arrondissement ID	Longitude	Latitude
1	2	−17.5251	14.7468
2	2	−17.5244	14.7474
3	2	−17.5226	14.7452

Table 2.3 Sample of bus stop data

Category	Line number	Bus station	Latitude	Longitude
Public network	2a	Daroukhane	14.7827	−17.3723
Public network	2b	Leclerc	14.6720	−17.4272
Public network	5a	Terminus Gudiawaye	14.7728	−17.3892
Public network	5b	Palais de Justice 1	14.6702	−17.443

2.2.3.2 Bus Data

Data from buses was gathered from Dakar's public transportation provider (DDD). DDD provides public transport services in two network categories: public and students, which were created to meet people's various travel needs.

Urban, suburban, and new circular lines (commuter, and urban) routes are covered by these two network user categories. From both the student and public networks, we discovered more than 33 routes (66 origin and destination stations). Some bus routes share stations, although their intermediate stops are different. As a result, the final number of origin-destination pairs is reduced to 24. Table 2.3 depicts a sample of DDD bus network data.

2.3   Methodology and Results

Our goal was to leverage large-scale mobile network data to estimate passenger travel demand so that strategic locations for potentially helpful transit routes and stations could be recommended. Origins and destinations must be identified in order to evaluate travel demand. Trips are usually performed in both directions. The commuting trip, in which trips are made back and forth between home and job, is a good example. As a result, there is travel demand in both areas. As a result, we projected a home-cell tower position for each subject based on the most frequently utilized cell tower location during the night (10 p.m.–7 a.m.) using our mobile network data. This home location detection method was first developed by Phithakkitnukoon et al. [43] who showed its reliability as they compared their result against the actual census data.

We also identified the top destinations for each subject based on the top five most visited cell tower locations (other than the subject's home cell tower location), i.e., the number of connections to each of the cell towers that the subject had used (or visited) was gathered for each subject, and the top five most connected cell tower locations (other than the subject's home cell tower location) were identified as a

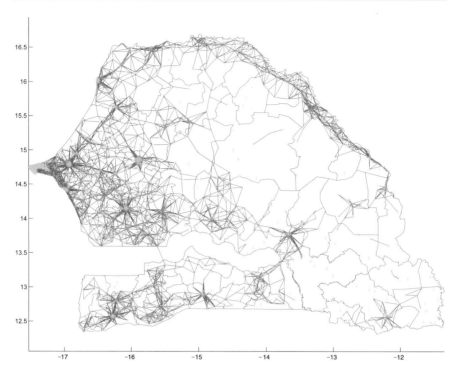

Fig. 2.2   Travel demand in Senegal, inferred from the mobile network data

result. As a result, we were able to connect each of the top five frequented places for each subject to the subject's home-cell tower position. These links (in red) are shown on the Senegal map below (Fig. 2.2), where cell tower locations are shown by green markers.

These links create cell-based flows, which can be interpreted as people's travel demand, and it can be seen that travel demand is relatively high in the Dakar region (Fig. 2.3), which is Senegal's most populated region, as well as along the coast of Thiès, which is one of the country's most touristic areas.

In the Saint-Louis and Matam regions, there is some significant demand along the north and north-eastern border regions. We compared our findings to the existing public transportation systems in Senegal in order to provide an informed recommendation about the public transportation systems. Despite the fact that Senegal's public transport data was limited, we were able to gather some information regarding public bus stop sites in Dakar, from which we were able to identify 24 origin-destination pairs of existing bus lines (shown in Fig. 2.4 below).

Each link between two places was made up of one or more cell-based flows (i.e., one or more individuals travel between the locations based on mobile network data), hence the flow density correlates to the level of travel demand in each link (between the two locations). Figure 2.5 shows a histogram of flow densities.

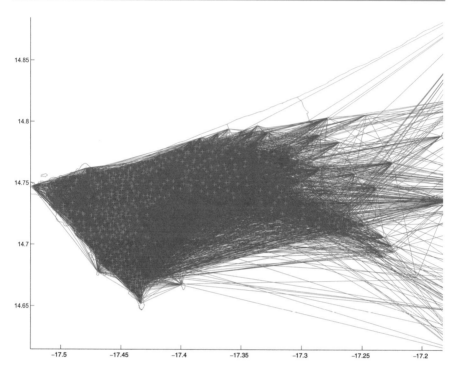

Fig. 2.3 Travel demand in Dakar, inferred from the mobile network data

The minimum flow is 1, the maximum flow is 17,479, the average flow is 60.39, and the standard deviation is 250.93. Intuitively, inferred travel demand varies by location, and in fact, a variety of factors influence travel demand, including land use, road network, socio-demographics, and so on.

We used the Dakar bus origin-destination pair link (Fig. 2.4 as the baseline) to map cell-based flows to public transportation demand. Our goal was to investigate how much of the inferred travel demand (cell-based flows) might be assembled into actual bus public transportation demand using Senegal's existing infrastructure. As a result, we began to re-construct the cell-based flows by adding cell-based flows one by one onto the map, from the highest flow density to the lowest, and then observed how cell-based flows build the bus routes; We started with a map of 24 bus links (origin-destination pairs) and incrementally overlaid cell-based flows onto the map from the largest to the smallest flow links, then kept track of where the bus and cell-based links intersected. We map-matched these locations by clustering cell towers to neighboring bus stations that are within 500 m, because the positions of the cell towers and bus stations are not exactly the same. We looked at a variety of closeness distances and discovered that this proximity distance of 500 m was a good choice for our investigation because it avoids overlapping and ensures that each bus station has at least one cell tower. We kept track of the percentage of actual bus routes assembled while we conducted our cell-based flow re-construction process (i.e.,

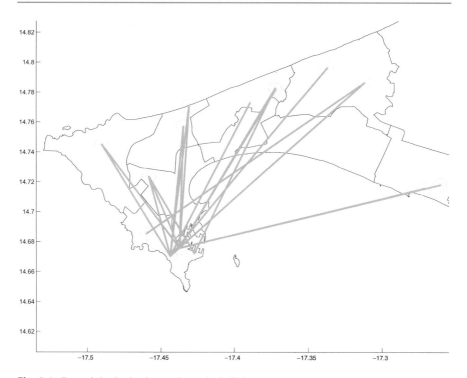

Fig. 2.4   Bus origin-destination station pairs in Dakar

how much it overlaps with bus routes), we discovered that just the top 22.70% cell-based flows were needed to fully assemble the existing bus routes (100%), as shown in Fig. 2.6.

Figures 2.7 and 2.8 depict the top 22.70% cell-based flows in Senegal and Dakar, respectively, where the flow density is represented by the thickness of the flow.

Using these top cell-based flows, which mimic Dakar's bus transportation network, to infer public transportation demand, it appears that there is significant demand in Dakar and emerging demand in other parts of the country, as well as a huge noticeable demand between Thiès and Diourbel.

We had inferred transit demands between cell tower locations, some of which were only a few meters apart. We further clustered the cell towers that are within 500 m of each other, using the same approach as with our bus station-based cell tower grouping, to approximate a prospective public transit station (i.e., stations are not too close to each other (within a walking distance, for example). The clustered cell towers in Dakar and Senegal are shown in Figs. 2.9 and 2.10 respectively.

We plotted the top 22.70% cell-based flows that resemble public transportation demand between possible public transportation stations using the clustered cell towers that resemble potential public transportation stations as shown in Fig. 2.11 (area of Dakar) and Fig. 2.12 (Senegal). We didn't include any current bus routes in

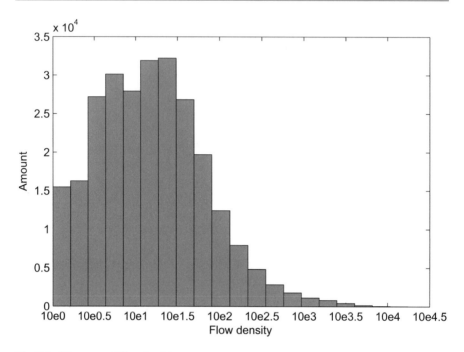

Fig. 2.5 Histogram of flow densities

these plots (Figs. 2.12 and 2.13) because we wanted to observe the travel demand in areas where public transportation is not available.

Based on our findings, we can make a sound suggestion for the creation of future transit routes that fulfil people's actual demands. Suggestions for the creation of additional transit routes to meet all referred transport demand (indicated in Fig. 2.12) may not be feasible due to the large government budget required. Our other possible suggestion is to identify potential transit routes that are desperately required, such as (1) top transit routes with flow densities larger than the average density, or (2) top transit routes with flow densities greater than half the average density (as shown in Figs. 2.13 and 2.14, respectively).

Option 1's suggested top inferred transit routes comprise a total of 47 potential routes, whilst option 2's top inferred routes include 234 suggested routes. The aforementioned conditions for using the average density are purely arbitrary. We just used these scenarios as examples to show how our inferred transport demand estimates could be used to aid decision-making in the planning and construction of new transportation routes.

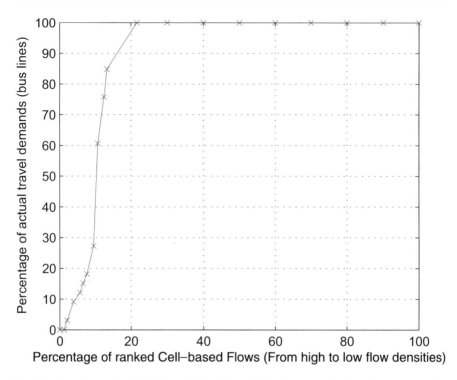

Fig. 2.6 Percentage of sorted cell-based flows (from high to low) based on flow density versus the percentage of actual bus origin-destination pair links assembled

2.4 Validation

Many factors influence whether we can estimate travel demand using mobile network data: the user must be active; the user must be a customer of SONATEL, the data provider; and calling plans, which can affect the number of samples obtained at each hour or day, and so on. As a result, validating our results against ground truth data is difficult. The fact that Senegalese transportation planners lack extensive travel surveys adds to the difficulty.

Nonetheless, we tried to validate our findings with census data [37]. The origin and destination locations of trips must be specified in order to assess travel demand. As a result, we projected a home-cell tower position for each subject based on the most frequently utilized cell tower location during the night using mobile network data. We tallied the number of respondents whose home-cell towers are located in each of Senegal's 14 governed regions, then matched our findings to official census statistics. Figure 2.15 displays a scatter plot of mobile phone users' predicted residential regional population against the actual regional population density derived from census data.

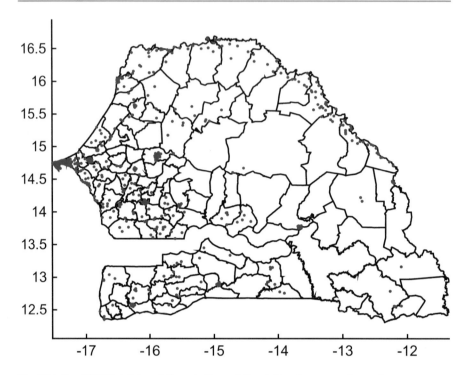

Fig. 2.7 Top 22.70% cell-based flows in Senegal (i.e., inferred transport demand)

The result indicates that the two estimates are comparable with a correlation coefficient of R = 0.85202. This result, we believe, confirms our approach. We did not show in our analysis if existing public transportation had enough capacity to handle the predicted ridership. Due to a lack of appropriate data, our attempt to analyze the capacity of transport lines was unsuccessful. For example, the buses do not adhere to DDD's actual timetables, making it difficult to get the required number of bus vehicles using a certain route at a specific time of day [44]; and there is no information on the number of paratransit vehicles going via a specific route. These details are necessary for calculating the capacity of a transit route in terms of the number of passengers per hour that can be transported. The widespread perception is that Senegalese people have inadequate transportation options to meet their mobility needs, and studies demonstrate that public transit in many areas is visibly underfunded. For example, in Senegal, walking, cycling, and other non-motorized transport systems support 45% of urban transportation needs [45]; and despite Dakar's population of more than 3 million, there is a lack of transportation, with only 25% of people's daily trips covered by motorized modes [46], and a qualitative survey conducted in Dakar also revealed a shortage in the supply side of the transportation system [47].

Our attempt to obtain local experts' input on the practicality of our findings was only partially successful. We were unable to obtain a complete trip survey in order to

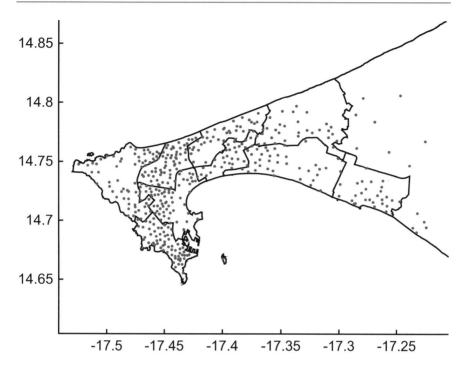

Fig. 2.8 Top 22.70% cell-based flows in Dakar region (i.e., inferred transport demand)

verify the expected travel demands. However, we obtained information about Senegal's road system and which parts of Dakar have the most traffic. For example, traffic on Dakar's east-west routes is extremely congested, especially during rush hour. The Senegalese government reacts by expanding the supply of road space by building more roads, such as a highway in Dakar that was opened to traffic in two phases: the Patte d'Oie-to-Pikine part in 2011 and the Pikine-to-Diamniadio section on August 1, 2013 [48]. Our investigation also revealed that there is a considerable demand for travel along this route.

More investigation was conducted to examine if the travel patterns predicted by mobile network data are consistent with urban movements on Senegal's primary road network. Roads account for over 90% of people and commodities movement in Senegal [49]. Senegal's present road network is divided into five levels: national highways, regional highways, department highways, urban highways, and classified tracks. National highways connect numerous administrative regions as well as adjacent states. Regional roads connect several departments within the same region. The remaining roadways serve as internal links between departments [49]. To further our analysis, we compiled individual visits made between Senegal's several districts (there are 123 districts). Figure 2.16a presents a qualitative picture of people's travel patterns between Senegalese districts, as inferred from mobile network data.

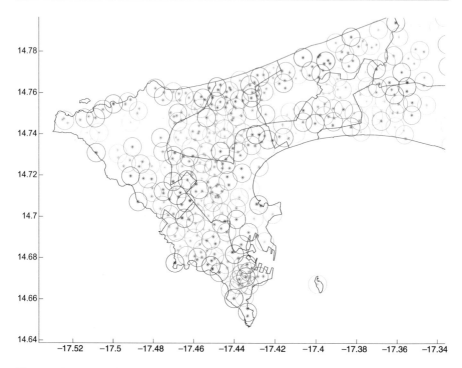

Fig. 2.9 Clustered cell towers in Dakar

Color differentiation is used to depict the aggregate flows of people in four groupings. The red color denotes strong flow (above the 75th percentile); the yellow color denotes flows between the 50th and 75th percentiles; and the green color denotes flows between the 25th and 50th percentiles, and the blue color stands for low flow (below the 25th percentile). The district centroids are shown by red markers. In Dakar, Thies, Saint-Louis, and Matam, there is a high concentration of movements within districts. Our conclusion clearly identifies the urban migration along the national road N2 from Dakar to Bakel, which is located on the northeast coast (with major hubs along the route: St-Louis, Richard Toll, Ndioum, and Orossogui).

This conclusion enables us to understand how people's travel patterns, as predicted from mobile network data, incorporate urban movements on Senegal's main road network as shown in Fig. 2.16b. The basic premise is that the amount and pattern of cellular traffic movement is increasing. Understanding the relationship between the intensity of urban movements and the current road network will aid in regulating urban dynamics.

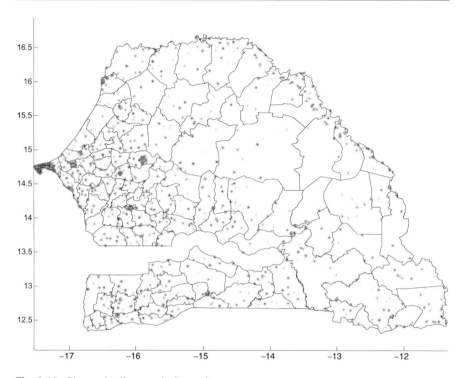

Fig. 2.10 Clustered cell towers in Senegal

2.5   Discussion of Potential Applications

The transport sector in Sub-Saharan Africa has been acknowledged as one of the critical components for growth, poverty alleviation, and sustainable human development by African planners and policymakers, economic experts, and international donors in the previous decade. A lot of efforts have been targeted towards interregional, interurban, and rural transportation sectors. Because the population of Sub-Saharan Africa has increased significantly and is likely to continue, the focus of Sub-Saharan Africa's transportation policy program is turning to enhancing urban mobility and accessibility [50, 51]. In the case of new bus route B-H, even though DDDs bus routes and E12 are already serving the Lopold Sdar Senghor International Airport, the addition of new route can be justified by limiting the number of intermediate stops to connect the downtown area with the Airport with shorter travel time.

  People's mobility in Dakar is aided in part by large-bus services and the paratransit system, which is the city's primary mode of transportation. Despite these services, some Dakar neighborhoods lack adequate public transportation, which our analysis clearly recognized and depicted (Figs. 2.13 and 2.14). We offer two options for transportation authorities in terms of developing possible public bus routes to meet

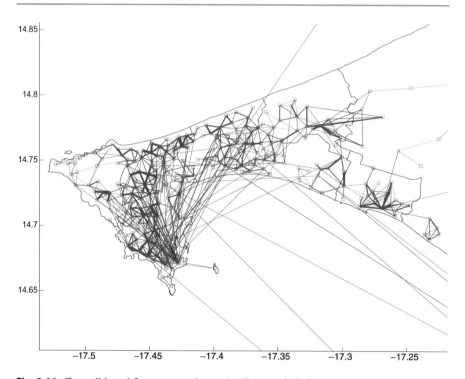

Fig. 2.11 Top cell-based flows across clustered cell towers in Dakar

the mobility demands of Dakar's underserved neighborhoods: (i) new public bus routes; and (ii) enhancing current public bus routes. These options might be prioritized based on budget constraints and the nature of transportation demand along a certain route.

As shown in Fig. 2.17, we found new transit route recommendations such as A-C, B-D, and B-H. The recommended routes connect the points A and B with the points C and D. Locations A and B are in Dakar's Plateau district (also known as downtown), which has a high concentration of office buildings, government buildings, retail, and hotels. Locations C and D are part of the Grand Yoff neighborhood, which is one of Dakar's most populous areas. While these regions are served by a robust urban public transportation system, our analysis revealed the need for more routes in areas where there is substantial demand for new bus routes. In the case of new bus route B-H, even though DDDs bus routes 8 and E12 are already serving the Lopold Sdar Senghor International Airport, the addition of new route can be justified by limiting the number of intermediate stops to connect the downtown area with the Airport with shorter travel time. Figure 2.17 also displays transit route options (B-F and E-G) to connect Dakar's central area (Plateau, Medina, Gueule Tapee Fass-Colobane, and Fann-Point E-Amitie) with Pracelles Assainies, one of the city's most populous areas.

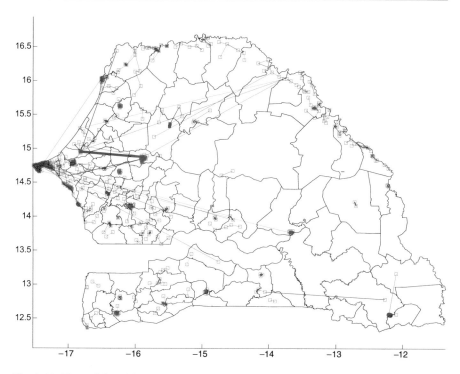

Fig. 2.12 Top cell-based flows across clustered cell towers in Senegal

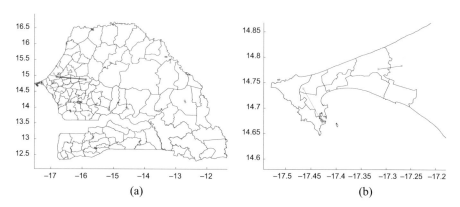

Fig. 2.13 Suggested 47 potential transit routes (top inferred routes with flow density greater than the average). (**a**) Senegal. (**b**) Dakar

We were able to determine a rather high flow density between the two destinations which supports the development of these new routes. The new routes can be established with a limited number of intermediate stops or as express versions of DDD's existing bus routes 1, 17, 23, and E4. The existing infrastructure, such as

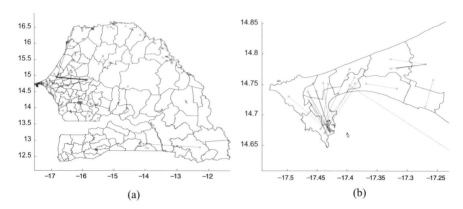

Fig. 2.14 Suggested 234 potential transit routes (top inferred routes with flow density greater than the average/2). (**a**) Senegal. (**b**) Dakar

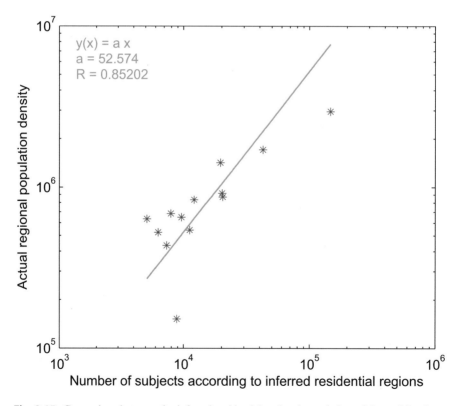

Fig. 2.15 Comparison between the inferred residential regional population of the mobile phone users and the actual regional population density obtained from the census data

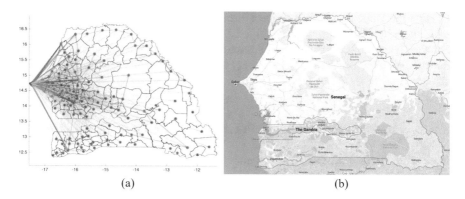

Fig. 2.16 Map of Senegal: (**a**) mobility of people between districts inferred from the mobile network data and (**b**) road network (Source: Google Maps)

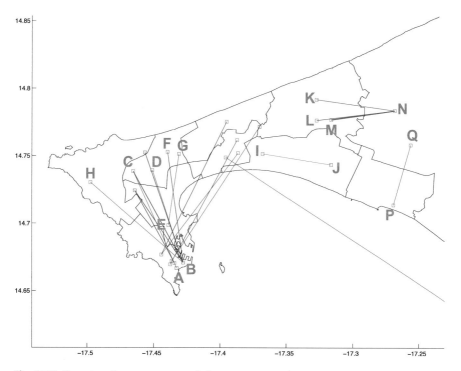

Fig. 2.17 Some transit route recommendations

terminals, can make introducing a new route on an existing corridor advantageous. This makes the implementation of new routes easier.

We also propose routes to connect the districts of Niayes and Bambilor (K-N, L-N, and M-N), as well as to improve public transportation ties between Rufisque and Bambilor districts (P-Q). The new routes can be launched as an extension of an existing bus route due to their modest distances. In the case of new bus route B-H, even though DDDs bus routes 8 and E12 are already serving the Lopold Sdar Senghor International Airport, the addition of new route can be justified by limiting the number of intermediate stops to connect the downtown area with the Airport with shorter travel time.

2.5.1 Improving the Current Practice of Urban Paratransit Service

In most situations, the paratransit system works in conjunction with or in addition to the bus system, which has been unable to expand in response to rising travel demand due to population expansion. Despite being criticized for its slowness in anticipating possible users at various points throughout the city, the Dakar paratransit system is designed to help the general population with mobility. As a result, the system is frequently faulty, and users have a hard time relying on it to plan their travel [52]. To remedy this issue, paratransit operators might use the proposed routes to reroute their services to the most in-demand regions. Furthermore, the paratransit system's use of small cars allows it to serve areas with low travel demand and increasingly dispersed trip-generating activities.

2.5.2 Providing Indicators for Potential High Order Public Transport Development

Senegalese cities today are dealing with rapid changes in population growth and urbanization, which constantly alter travel requirements. Authorities are looking for ways to improve public transportation's quality, reliability, and coverage. One option is to improve public transportation by bolstering the operation of existing paratransit and large-bus services, while another is to create a hybrid system that combines paratransit and high-order public transportation [53]. Several local governments in developing nations have taken this initiative and suggested a BRT system [54]. CETUD tasked SAFEGE/SCE with conducting engineering studies for a pilot dedicated-lane BRT system in Dakar [55]. Estimating and forecasting public transportation demand using traditional methods, which require massive data collection work such as expensive travel surveys, demographic and land use characteristics [56], is one of the major issues associated with this type of expansion. In this sense, transportation planners can use our analysis methodology to better understand people's mobility patterns, and the recommended transit routes can be included in a future high-order transportation system.

2.5.3 Cost-Effective Transport Planning Approach

The early steps in public transportation planning, such as estimating travel demand using classic transportation models, necessitate a significant amount of data collecting. Furthermore, the time it takes to estimate a model and forecast demand can be years, during which time the dynamics of urban activities, transportation systems, and relevant policies can change, necessitating new data collection and modeling efforts [57]. Our research uses the large amount of mobile network data available in Senegal to show how transportation planners can sense people's movements more regularly, at a lower cost, and on a larger scale, especially in situations where relevant data is unavailable or poorly supplied. This may encourage transport planners in underdeveloped nations to move away from the costly and time-consuming traditional transportation planning approach and toward a more data-driven planning method in which individuals are part of the sensing infrastructure.

2.6 Conclusion

Due to the limited budget available for transport data collection in developing countries, planners face significant challenges in capturing current and future mobility trends. Data from alternative sources is now becoming available and can be used to sense the movement of people in cities on a more regular basis, at a lower cost, and on a larger scale. It's especially useful and necessary when relevant data isn't available or is provided in a poor manner. In today's cities, cellular networks are ubiquitous, and the high mobile phone penetration rate in developing countries allows planners to opportunistically sense people's presence and mobility at fine-grained resolution. We argue that, despite the low state of development of transportation planning practice in developing countries, they can benefit from a decade of research on using mobile network data to create a more efficient public transportation system without having to go through the costly structure that developed countries have set up for this purpose.

In this chapter, we discuss a methodology to extract people's movement patterns using their mobile phone usage as a proxy. The findings allow transportation planners (public transportation operators) to assess their public transportation services and efficiently build new transit routes or expand existing routes to meet consumers' mobility needs by increasing bus route coverage and accessibility. Senegal was utilized as a case study country to demonstrate our methodology. A large number of stakeholders will benefit from this research. For example, public transportation companies would be able to better analyze transportation demand and tailor their service offers based on the extracted mobility patterns. City officials will have greater opportunity to tackle social and economic problems with better knowledge on people's mobility patterns and will be able to make more informed judgments about future sustainable urban infrastructure development to better serve inhabitants. Above all, this research contributes to creation of a more sustainable urban mobility system by maximizing the use of low-cost mobile network data and enhancing coverage and accessibility and efficiency of the public transport systems. Despite the importance of our findings, there are some significant limitations to the study. In principle, notwithstanding SONATEL's 61% share of the Senegalese market, this analysis is applicable to only SONATEL subscribers. To link the extracted movements and flows to all Senegalese citizens, we must assume that our sample is reflective of the country's current population distribution. Other factors that may have an impact on the results of our analysis include: the number of mobile devices (including Subscriber Identity Module (SIM) cards) carried by each individual; and calling plans, which might affect the quantity of samples collected at each hour or day. Nonetheless, a study in Kenya found that estimates of mobility based on mobile network data are resilient to significant biases in phone ownership across user groups [57].

People's travel patterns informed the transportation route recommendations. In addition to travel pattern data, it's also critical to consider whether the new route will reduce travel time or increase ridership, as well as whether out-of-vehicle times like access, waiting, and transfer will be reduced. Our study did not take into account the newer information, and we focused our investigation on making recommendations for convenient connections between users' origin and destination in high-demand locations. However, it is important to remember that bus routes designed to reduce travel time may not meet the requirements for greatest coverage and accessibility.

We distributed all of the transportation demand to the public transportation networks that were available. This is attributable to the fact that public transportation accounts for more than 85% of all journeys in Senegal's urban areas [45]. Our approach complements the traditional multi-stage travel demand forecasting approach. One of the ultimate goals is to give transportation planners, particularly in developing nations, an option to consider in the absence of comprehensive transportation data when estimating ridership and planning transportation networks.

Our route suggestions are primarily based on traffic flows between the origin and destination points (terminus locations). Future studies will need to take into account the presence of major trip producers along the transit route while deciding on intermediate stations. We were unable to gather complete information about Dakar's urban transit system. We used transit network data from DDD to validate our findings, including bus routes, network type, and origin and destination bus stop locations. This analysis can be further explored by including more relevant data, such as existing transit route capacity, bus ridership data, and more information about paratransit services as they become available. Furthermore, we only looked at people's mobility patterns between their homes and the top five destination sites, ignoring any intermediate activities they might engage in. We do realize, however, the need of combining intermediary actions of each subject, which our study currently lacks. When more data from the research region becomes available, this will be worth an investigation.

References

1. United Nations. United Nations population fund (UNFPA). 2014. https://www.unfpa.org/about-us.

2. African Development Bank. Tracking Africa's progress in figures. 2014. http://www.afdb.org/fileadmin/uploads/afdb/Documents/Publications/Tracking_Africa's_Progress_in_Figures.pdf.
3. Rodrigue JP, Comtois C, Slack B. Urban transportation. In: The geography of transport systems. 1st ed. New York; 2006. p. 191–2.
4. Kirby RF, Bhatt KU, Kemp MA, McGillivary RG, Wohl M, Institute U. Para transit: neglected options for urban mobility. Washington: Urban Institute; 1974.
5. Behrens R, Mccormick D, Mfinanga D. An evaluation of policy approaches to upgrading and integrating paratransit in African urban public transport systems: results of the first round of a Delphi survey. In: Conf. CODATU XV; 2012. p. 1–24.
6. Transport SA, Program P, Bank TW. Urban mobility in three cities. Addis Ababa. 2002;70.
7. Williams B. Sustainable urban transport in Africa: issues and challenges. 2011. https://www.itu.int/en/ITU-%0AD/Statistics/Documents/facts/ICTFactsFigures2013-e.pdf.
8. ITU, International Telecommunication Union (ITU). 2013. https://www.itu.int/en/ITU-%0AD/Statistics/Documents/facts/ICTFactsFigures2013-e.pdf.
9. Boyle D. Fixed-route transit ridership forecasting and service planning methods. Transit Coop Res Prog. 2006. http://onlinepubs.trb.org/onlinepubs/tcrp/tcrp_syn_66.pdf.
10. Big data helps city of Dublin improve its public bus transportation network and reduce congestion. Int Business Machines (IBM). 2013.
11. Wang W, Attanucci JP, Wilson NHM. Bus passenger origin-destination estimation and related analyses using automated data collection systems. J Public Transp. 2011;14(4):131–50. https://doi.org/10.5038/2375-0901.14.4.7.
12. Foell S, Kortuem G, Rawassizadeh R, Phithakkitnukoon S, Veloso M, Bento C. Mining temporal patterns of transport behaviour for predicting future transport usage. In: Proceedings of the 2013 ACM conference on pervasive and ubiquitous computing adjunct publication; 2013. https://doi.org/10.1145/2494091.2497354.
13. Foell S, Phithakkitnukoon S, Kortuem G, Veloso M, Bento C. Catch me if you can: predicting mobility patterns of public transport users. In: 17th international IEEE conference on intelligent transportation systems (ITSC); 2014. https://doi.org/10.1109/ITSC.2014.6957997.
14. Bar-Gera H. Evaluation of a cellular phone-based system for measurements of traffic speeds and travel times: a case study from Israel. Transp Res Part C Emerg Technol. 2007. https://doi.org/10.1016/j.trc.2007.06.003.
15. Wang P, Hunter T, Bayen AM, Schechtner K, González MC. Understanding road usage patterns in urban areas. Sci Rep. 2012;2(1). https://doi.org/10.1038/srep01001.
16. Demissie MG, Correia GH de A, Bento C. Exploring cellular network handover information for urban mobility analysis. J Transp Geogr. 2013. https://doi.org/10.1016/j.jtrangeo.2013.06.016.
17. Demissie MG. Combining datasets from multiple sources for urban and transportation planning. Coimbra Univ.; 2014.
18. Demissie MG, de Almeida Correia GH, Bento C. Intelligent road traffic status detection system through cellular networks handover information: an exploratory study. Transp Res Part C Emerg Technol. 2013. https://doi.org/10.1016/j.trc.2013.03.010.
19. Hongsakham W, Pattara-atikom W, Peachavanish R. Estimating road traffic congestion from cellular handoff information using cell-based neural networks and K-means clustering. In: 2008 5th international conference on electrical engineering/electronics, computer, telecommunications and information technology; 2008. https://doi.org/10.1109/ECTICON.2008.4600361.
20. Becker RA, et al. Route classification using cellular handoff patterns. In: Proceedings of the 13th international conference on ubiquitous computing; 2011. p. 123–32. https://doi.org/10.1145/2030112.2030130.
21. Herrera JC, Work DB, Herring R, Ban XJ, Jacobson Q, Bayen AM. Evaluation of traffic data obtained via GPS-enabled mobile phones: The Mobile Century field experiment. Transp Res Part C Emerg Technol. 2010;18(4). https://doi.org/10.1016/j.trc.2009.10.006.

22. Puntumapon K, Pattara-atikom W. Classification of cellular phone mobility using naive Bayes model. In: VTC spring 2008 – IEEE vehicular technology conference; 2008. p. 3021–5. https://doi.org/10.1109/VETECS.2008.324.
23. Çolak S, Alexander LP, Alvim BG, Mehndiratta SR, González MC. Analyzing cell phone location data for urban travel. Transp Res Rec J Transp Res Board. 2015;2526(1):1–17. https://doi.org/10.3141/2526-14.
24. Alexander L, Jiang S, Murga M, González MC. Origin–destination trips by purpose and time of day inferred from mobile phone data. Transp Res Part C Emerg Technol. 2015;58:240–50. https://doi.org/10.1016/j.trc.2015.02.018.
25. Toole JL, Colak S, Sturt B, Alexander LP, Evsukoff A, González MC. The path most traveled: travel demand estimation using big data resources. Transp Res Part C Emerg Technol. 2015;58:161–428. https://doi.org/10.1016/j.trc.2015.04.022.
26. Alexander MGL. Assessing the impact of real-time ridesharing on urban traffic using mobile phone data. In: Proc. UrbComp; 2015. p. 1–9.
27. Jiang S, Ferreira J, Gonzalez MC. Activity-based human mobility patterns inferred from mobile phone data: a case study of Singapore. IEEE Trans Big Data. 2016;3(2):208–19. https://doi.org/10.1109/tbdata.2016.2631141.
28. Caceres N, Wideberg JP, Benitez FG. Deriving origin–destination data from a mobile phone network. IET Intell Transp Syst. 2007;1(1):15–26. https://doi.org/10.1049/iet-its:20060020.
29. Calabrese F, Di Lorenzo G, Liu L, Ratti C. Estimating origin-destination flows using mobile phone location data. IEEE Pervasive Comput. 2011. https://doi.org/10.1109/MPRV.2011.41.
30. Iqbal MS, Choudhury CF, Wang P, González MC. Development of origin–destination matrices using mobile phone call data. Transp Res Part C Emerg Technol. 2014;40:6374. https://doi.org/10.1016/j.trc.2014.01.002.
31. Demissie MG, Correia G, Bento C. Analysis of the pattern and intensity of urban activities through aggregate cellphone usage. Transp A Transp Sci. 2015. https://doi.org/10.1080/23249935.2015.1019591.
32. Soto V, Frias-Martinez E. Robust land use characterization of urban landscapes using cell phone data. In: Proceedings of the 1st workshop on pervasive urban applications, in conjunction with 9th int. conf. pervasive computing, vol. 9; 2011.
33. Toole JL, Ulm M, González MC, Bauer D. Inferring land use from mobile phone activity. In: Proceedings of the ACM SIGKDD international workshop on urban computing; 2012. https://doi.org/10.1145/2346496.2346498.
34. Ahas R, Silm S, Järv O, Saluveer E, Tiru M. Using mobile positioning data to model locations meaningful to users of mobile phones. J Urban Technol. 2010;17(1):3–27. https://doi.org/10.1080/10630731003597306.
35. Csáji BC, et al. Exploring the mobility of mobile phone users. Phys A Stat Mech its Appl. 2013;392(6):1459–73. https://doi.org/10.1016/j.physa.2012.11.040.
36. Isaacman S, et al. Identifying important places in people's lives from cellular network data. In: International conference on pervasive computing. Berlin, Heidelberg: Springer; 2011. p. 133–51.
37. Berlingerio M, Calabrese F, Di Lorenzo G, Nair R, Pinelli F, Sbodio ML. AllAboard: a system for exploring urban mobility and optimizing public transport using cellphone data. In: Joint European conference on machine learning and knowledge discovery in databases. Berlin, Heidelberg: Springer; 2013. p. 663–6.
38. Geohive. Senegal census data. 2014. http://www.geohive.com/cntry/senegal_ext.aspx.
39. Trans-Africa Consortium. Overview of public transport in Sub-Saharan Africa. 2008. https://citeseerx.ist.psu.edu/viewdoc/download?doi=10.1.1.179.2865&rep=rep1&type=pdf.
40. World Bank. Senegal transport and urban mobility project project: appraisal document. World Bank; 2010. https://documents1.worldbank.org/curated/en/825191468305338767/pdf/537740PAD0P1011y100IDA1R20101014511.pdf.
41. Deloitte and GSMA. Sub-Saharan Africa mobile observatory. 2012. http://www.gsma.com/publicpolicy/wp-content/uploads/2012/03/SSA_FullReport_v6.1_clean.pdf.

42. de Montjoye Y-A, Smoreda Z, Trinquart R, Ziemlicki C, Blondel VD. D4D-Senegal: the second mobile phone data for development challenge. arXiv. 2014;1407(4885):1–11. [Online]. https://arxiv.org/pdf/1407.4885.pdf

43. Phithakkitnukoon S, Smoreda Z, Olivier P. Socio-geography of human mobility: a study using longitudinal mobile phone data. PLoS One. 2012;7(6):e39253. https://doi.org/10.1371/journal.pone.0039253.

44. Godard X. Dakar experience in bus reform. World Bank Group; 2005.

45. Road travel report: Senegal. In: Alberta serious incident response team (ASIRT); 2009.

46. Godard X. Urban transport reform in Dakar, lessons from 15 years experience, the search of complementarity between bus and minibus operators. In: Proc. 11th world conf. transp. res; 2007. p. 1–16.

47. Bipe/Ter. Impact Social de la Crise ne du système de déplacements Dakar, Paris. 2000.

48. EIFFAGE. Projets, ouvrages & réalisations Eiffage. 2013. https://www.eiffage.com/ouvrages/autoroute-de-lavenir.html.

49. DLCA. Senegal road network. 2013. https://dlca.logcluster.org/display/public/DLCA/2.3+Senegal+Road+Assessment. Accessed 15 Oct 2015.

50. Africa Transp. Policy program (SSATP), preparation of third development plan. Urban transport—mobility and accessibility cluster. In: Africa transp. policy program (SSATP); 2012. http://www.ssatp.org/sites/ssatp/files/publications/HTML/Conferences/Addis12/Apendices/03- Urban-Mobility-Addis_EN.pdf.

51. Gorham R. Sustainable mobility and accessibility in urban areas of Africa. 2015.

52. Agyemang W. Measurement of service quality of 'Trotro' as public transportation in Ghana: a case study of the city of Kumasi. 2013.

53. Salazar Ferro P, Behrens R, Wilkinson P. Hybrid urban transport systems in developing countries: portents and prospects. Res Transp Econ. 2013;39(1). https://doi.org/10.1016/j.retrec.2012.06.004.

54. Wright L, Hook W. Bus rapid transit: planning guide. Inst Transp Develop Policy. 2007.

55. World Bank. Dakar bus rapid transit pilot project. https://projects.worldbank.org/en/projects-operations/project-detail/P156186.

56. Mullen P. Estimating the demand for urban bus travel. Transportation. 1975;4(3):231–52. https://doi.org/10.1007/BF00153577.

57. Wesolowski A, Eagle N, Noor AM, Snow RW, Buckee CO. The impact of biases in mobile phone ownership on estimates of human mobility. J R Soc Interface. 2013;10(81). https://doi.org/10.1098/rsif.2012.0986.

Modeling Trip Distribution Using Mobile Phone CDR Data

3

Abstract

Creation of a trip distribution model necessitates a large amount of data collection, such as costly travel surveys to determine trip makers' origin and destination zones. Intrazonal trips are frequently overlooked when developing trip distribution models due to the difficulties in calculating their travel expenses. Ignoring intra-zonal travels, on the other hand, leads to inaccurate model estimates, especially when the zone size is large and the number of intra-zonal visits is high. This is especially relevant given the increased interest worldwide in making cities more walkable and bikeable, where intra-zonal travel accounts for a large portion of those journeys. We employ mobile phone network data to generate country-wide mobility trends in this chapter. For 123-district level traffic analysis zones, a set of doubly constrained trip distribution models that integrate intra-zonal trips is estimated. We then examine two methods for calculating intra-zonal travel expenses based on trip distance. The average intra-zonal trip distances determined from the two techniques yields differing levels of sensitivity to the distance-decay effect, according to our analysis. This demonstrates that in the absence of intrazonal trips, model estimation produces erroneous results. The content discussed in this chapter is inspired our original work by Demissie et al. (IEEE Trans Intell Transp Syst. 2019;20(7):2605–17; 13th international conference on electrical engineering/electronics, computer, telecommunications and information technology (ECTI-CON); 2016. p. 1–6).

Keywords

Call detail records · Trip distribution modeling · Origin-destination matrix · Gravity model · Log-linear model · Intra-zonal trips

S. Phithakkitnukoon, *Urban Informatics Using Mobile Network Data*,
https://doi.org/10.1007/978-981-19-6714-6_3

3.1 Motivation and State of the Art

Travel flow estimation at various geographical and temporal dimensions is a recurring topic in various fields of study. Aside from individuals moving from one area to another, there are many different types of flows, such as raw material or goods distribution, money flows in economics, or particle flows in physics. People moving from one location to another can be classified according to their temporal and spatial characteristics. Johnson [1] classified short-distance and short-duration flows as commuting to work/school/shopping, and long-distance and long-duration flows as internal/global migration.

Trip distribution is one of the main stages of the traditional four-step transportation planning model when it comes to estimating travel demands. This step represents the number of journeys made between trip origins and destinations, as well as the pattern of trip making activity. Various types of trip distribution models have been created over time [2, 3]. Some of the most basic models, such as the growth-factor model, are suitable for short-term studies where no large changes in the transportation network are expected. There are, however, situations that induce fluctuations in the cost of the transportation network. The gravity model is one of the most well-known models for long-term strategic analyses. When major changes in the transportation network occur, this model responds better to changes in the trip pattern.

Travel surveys have traditionally been used to collect mobility data. Large-scale travel surveys have the advantage of giving precise information, but they are typically costly and unavailable to transportation planners and policymakers [4]. The sample size of small-scale travel surveys, such as roadside interviews or public transportation questionnaires, is typically modest. The use of big and small scale travel surveys in combination does not capture enough origin-destination (OD) pairs from all traffic analysis zones [2]. Furthermore, planners frequently overlook intra-zonal journeys due to the difficulties of monitoring and forecasting their travel costs. When intra-zonal trips are ignored during model estimation, the parameter estimate becomes skewed [5]. Intrazonal trips are underestimated, resulting in much higher estimates of journeys to other zones [6, 7]. At the local level, predicted trip distribution models may be less susceptible to traffic congestion and pollution [8]. In addition, non-motorized trips account for a large share of intra-zonal travel (e.g. active modes such as walking and cycling). Planning for active transportation modes has become one of the most important aspects of transportation and urban planning in recent years. For public health research [9], bicycle and pedestrian safety studies [10, 11], and feasibility studies for active modes infrastructure enhancements [12], practitioners and policymakers need precise information on non-motorized trips volumes.

In this chapter, we present a computationally feasible way for deriving a country's mobility patterns from mobile phone network data or Call Detail Records (CDRs), as well as insights into how our methods can be used in the absence of extensive mobility survey data. In particular, we investigate the use of CDRs data to derive more realistic OD trips, then investigate the possibility of using mobile phone data to measure intra-zonal trip distances and the impact of the presence or absence of intrazonal trips on model estimation, and lastly investigate the interaction between the statistical and entropy maximising spatial interaction models.

The data necessary for travel demand modeling includes information about the origin and destination of trips. Traditionally, this information has been gathered mostly through travel surveys, which require respondents to keep track of where they are traveling to and from during specific observation times, the forms of transportation they use, the purpose of their journey, their routes, and so on. Despite the fact that it provides extensive information, performing a travel survey takes time, and large-scale surveys are time-consuming and costly. Recently, the emphasis has switched to the use of opportunistic datasets derived from a variety of sources that can provide insights into the spatial distribution and temporal evolution of citizen and vehicle movements. Transport planners now have new tools for assessing people and vehicle flows, as well as examining their mobility demands, thanks to the analysis of mobile phone data [13], GPS data [14–16], and open and crowd-sourced data [17, 18].

Since the early 2000s, the use of cellular network data for the development of large-scale mobility sensing has been investigated [19]. The data have been used to explore various aspects of transport issues: large-scale urban sensing [20, 21]; traffic parameter estimate [22–24]; origin-destination trip estimation [25–27]; land use inference [28], [29]; travel demand estimation [30–33]. Despite the data's versatility in supporting a variety of studies, it has two major drawbacks [34]: (i) CDR is sparse in time because it is only acquired when a device is engaged in a voice call or short message service; and (ii) CDR is coarse in space because the location record is made available at the granularity of cell tower service coverage. Depending on how the CDR data will be used, these limitations will have a different impact. CDR data, for example, may not provide detailed information in terms of providing absolute traffic volume counts and distinguishing between routes for the purposes of estimating travel time, travel speed, and traffic volume [23]. The CDR data used in this work, on the other hand, is nevertheless helpful for our analysis, which is at a scale where a lack of precision is acceptable.

Previous research has shown that cell phone data can be utilized to update OD flow estimates more often, cutting down on the time it takes to calculate OD flows using standard methods. In contrast to data gathered through traditional surveys, this approach can be repeated with freshly accessible datasets at a lower cost and with a higher frequency [25]. The veracity of OD flow obtained from mobile phone data, on the other hand, is debatable, particularly when sample and penetration rates are insufficient [26, 35]. Calabrese et al. [26] and Csáji et al. [36] use the fit of the data to a gravity model to determine the accuracy of the estimated OD flows. Jiang et al. [37] and Schneider et al. [38] employed household surveys to validate the authenticity of individual users' mobile phone trajectories. The use of mobile phone data to derive trip distribution for different modes, trip reasons, and time of day [30, 31, 39–41] is one of the most recent advances in the field.

Several trip distribution models have been developed using data from mobile phones [36, 42–44]. Wang et al. [44] optimized Senegal's national and regional road network using the output of a novel type of gravity model. One of the study's shortcomings is that the model provides relative traffic volume, therefore capacity cannot be used as a road design feature. Furthermore, the intra-zonal journeys were not included in this analysis, which could have resulted in a major underestimation of congestion [45]. Venigalla et al. [8] cited a lack of networks as the primary reason for ignoring intrazonal trips. Another problem is the difficulty of calculating the cost of intra-zonal journeys [5, 46, 47].

The centroid-to-centroid distance between the origin and destination zones is commonly used to evaluate travel costs. The centroid-to-centroid model, on the other hand, can result in a zero-distance separation of intra-zonal flow, which is always positive in reality [47]. The United States made one of the first attempts to determine intra-zonal journey distance. Using the average distance between the centroid of a zone and the centroids of adjacent zones divided by two [48], the Department of Commerce calculated the average distance between the centroid of a zone and the centroids of adjacent zones. Venigalla et al. [8] used a similar method to calculate the inter-zonal journey distance as half the distance between the zone's centroid and the closest zone's centroid. Batty [49] uses the assumption that the traffic analysis zone is circular and that the population is distributed evenly to calculate the inter-zonal trip distance, which takes the shape of Eq. (3.1).

$$C_{ii} = \frac{r_i}{\sqrt{2}}, \tag{3.1}$$

where C_{ii} is the intra-zonal trip distance and r_i is the zone radius.

Intra-zonal trip distance estimation was also approached by Batty [49] as a problem of determining the mean trip length within a traffic analysis zone. In this case, the most appropriate method of estimating intra-zonal trip distance is to disaggregate trip and distance data within each individual zone. The C_{ii} can be determined by dividing each traffic analysis zone into x origin subzones and y destination subzones as follows.

$$C_{ii} = \frac{\sum_x \sum_y T_{xy} C_{xy}}{\sum_x \sum_y T_{xy}}, \tag{3.2}$$

where, T_{xy} is the number of trips between origin and destination subzones x and y (x, y $\in i$), and C_{xy} is the distance between them. The trip distance associated with intra-zonal trips, on the other hand, is frequently unavailable, making this method difficult to use. The intra-zonal trip distance is computed using both approaches in this study. Cellular network base stations inside a specified traffic analysis zone are employed as the center of subzones x and y in the second technique.

3.2   Methodology

Five major steps required to develop a spatial interaction model using mobile phone data is depicted in Fig. 3.1, where (A) we want to know if users are staying in a given zone, doing something in it, or passing through it en route to their destination; (B) home and work zones are inferred for each user based on the duration and most frequently used cell tower locations during the night and day; (C) OD flows are estimated using the information from stages A and B; (D) Data expansion is performed using a population expansion factor that connects sample users to the population; and (E) a set of doubly constrained trip distribution models is developed.

3.2.1   Case Study Region and Dataset

Senegal is used as a case study location to show how this analysis was carried out. Senegal had an estimated population of 13,508,715 in 2013. The country is organized into 14 regions, 45 departments, and 123 arrondissements (districts) [50]. Because our analysis is done at the district level, there are a total of 123 traffic analysis zones (TAZ).

There were 1666 cell towers in total located across the TAZ. TAZ that is in densely inhabited areas (e.g., Dakar, Thies, Saint-Louis, etc.) has more cell towers serving a smaller area and fewer cell towers covering a wider geographic area in sparsely populated areas. The average number of cell towers per TAZ is 14, with the first, second, and third quartiles having 6, 9, and 15 respectively. Between January 7 and January 20, 2013, we used anonymized CDRs from Senegalese mobile phone users for a 2-week period (voice and short message service data). Within the D4D Senegal Challenge framework, SONATEL and Orange make the data public. SONATEL had a 61% market share in Senegal in 2012 [51]. The cell tower granularity was used to obtain the mobile phone records, and each record contains an anonymized unique user ID as well as the date and time of the call.

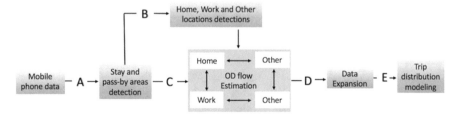

Fig. 3.1 Flow chart of methodology

3.2.2 Stay and Pass-by Area Identification

By tracing the trajectories between two cell towers, a simple method for determining the movement of a mobile phone user can be implemented. A trip can be identified, for example, when the user connects to the cellular network using the nearest cell tower at the time and then reconnects using a different cell tower nearby. However, a trip inferred in this manner may not necessarily reflect the user's actual mobility pattern. If a user connects to the cellular network in a certain cell tower region while driving to his or her destination, that cell tower location should not be designated as the trip's origin or destination. As a result, such a method may incorrectly represent a long trip as a sequence of short trips. Previous research by Zheng et al. [14] and Hariharan and Toyama [15] established a sound methodology for determining a stay location, or a site where a user stayed for a period of time, from GPS trajectories. Alexander et al. [30] used CDR data with triangulated coordinates to assess the length of time a person stayed in a certain area. We used CDR data in this analysis, which has a lesser spatial resolution because locations are represented by cell towers. We employ a similar method to isolate traces that can be used to deduce the users' origin/destination zones. However, due to the nature of the CDR data in our analysis, we adopt some of Wang et al. [44]'s recommendations.

For each mobile phone user x, we arranged the observed data (M_x) in consecutive order along the date and time of the day. $M_x = (m_x(1), m_x(2), m_x(3), \ldots, m_x(n_x))$. Where, $m_x(k) = (d_x(k), t_x(k), l_x(k))$ for $k = 1, \ldots, n_x$ and $d_x(k), t_x(k), l_x(k)$ are the day, time, and cell tower location of the kth observation of the user x. Then, the consecutive traces at the same tower location of each user are grouped together. For each user x, the rth group of traces the user has is at cell tower location L_{xr}, and the first connection time at L_{xr} is FT_{xr} and the last connection time at L_{xr} is LT_{xr}. The amount of time user x spent at L_{xr} can be computed as $LT_{xr} - FT_{xr}$. We impose a criterion on the data to categorize the presence of user x at a given cell tower location as stay or pass-by. Thus, if $LT_{xr} - FT_{xr} < 10$ min, the rth group of traces that the user x had should be removed from the traces since the cell tower location L_{xr} is regarded as a transient location passed by the user heading to the destination/origin. In this algorithm, $x \in \{1, 2, 3, \ldots, 319{,}508\}$, $r \in \{1, 2, 3, \ldots\}$, $L_{xr} \in \{1, 2, 3, \ldots, 1666\}$ and at the same day $L_{xr} \neq L_{x(r+1)}$, FT_{xr} and LT_{xr} are in the form of $yyyy - mm - dd\ hh : mm : ss$.

Data filtering was the first step in identifying stay and pass-by areas. The original dataset contained over 9 million unique aliased mobile phone numbers, and Orange Labs further processed the data to retain users who met the following two criteria: (i) users who had more than 75% days with interactions per given period; and (ii) users who had less than 1000 interactions per week. These precautions are taken to ensure a high sampling rate by including only active users in our sample, as well as to filter out noise caused by the extraordinarily large number of mobile phone exchanges, which are assumed to be machines or shared phones [52].

We looked at the remaining 44.4 million mobile phone connections from 319,508 people over the course of 2 weeks, from January 7 to January 20, 2013. The data provides first, second and third quartiles of 51, 86, and 157 phone interactions per

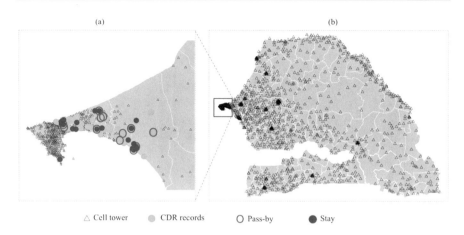

Fig. 3.2 Stay and pass-by areas of a sampled user (**a**). Study region (**b**)

person respectively. Figure 3.2 uses the mobile phone records of a randomly selected user to show the stay and pass-by locations. Figure 3.2a displays 462 mobile phone records from the sampled user, which are scattered among eight Senegalese districts (solid green circles). Figure 3.2a also depicts 68 remain (solid blue circles) and 16 pass-by (open red circles) places derived from the 462 mobile phone records collected over a 2-week period.

The data points are reduced from the original 44.4 million observations to 7.5 million stay locations (the number of times where the time duration between consecutive traces/calls is more than 10 min) after the consecutive traces at the same tower location of each user are grouped together and consecutive traces with time duration of less than 10 min are eliminated. These stay places are linked to cell tower sites, which can be aggregated across Senegal's 123 districts. Figure 3.3a shows the average stay locations per user of 19.11, with first, second, and third quartiles of 9, 14, and 23 over a 2-week period. The number of stay spots aggregated by Senegal districts is shown in Fig. 3.3b. Each sample user visited 2.08 districts on average, with the first, second, and third quartiles visiting 1, 2, and 3 districts throughout a 2-week period.

3.2.3 Significant Location Detection

A person's significant locations can be determined by ranking the frequency and duration of visits to each location [53]. The significant locations in this study were identified using a combined measure of the number of calls (frequency) and stay time (duration). The number of times each cell tower is contacted during the night (10 p. m.–7 a.m.) is measured for each individual user to determine their home location. The cell tower with the most interactions is then designated as the home location. The analysis excludes users who have had fewer interactions at their home places.

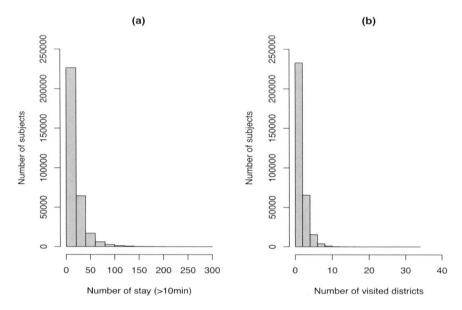

Fig. 3.3 Number of stays per person (**a**); and number of visited districts per person (**b**)

When a user has a similar number of interactions at two or more cell tower locations, the following procedure is used: (i) stay point arrival and departure times are used to calculate the amount of time the individual user spent in each cell tower location; (ii) the cell tower with the longer stay time is chosen as the home location. To eliminate bias from normal non-working days, we first filtered out calls on weekends (Saturday and Sundays) to establish the work location. Then, during the day, we followed the aforementioned procedures (8 a.m.–7 p.m.).

We acknowledge that our assumptions are basic and that we may not be able to precisely pinpoint the user's home and work location. The term 'work location,' for example, can be misleading; it's likely that some of these devices were used by schoolchildren or non-workers who went somewhere during the day. Our study does not take into consideration the number of people who work overnight.

Census data was used to validate the home location identification method in previous studies [32, 53]. These studies looked at the relationship between the inferred residential population of mobile phone users and the actual population based on census data, and discovered a strong correlation. Additional analysis, including the distribution of the number of hours people spend away from their workplaces, shows that the work location estimation method is still valid (Fig. 3.4).

Only users with inferred home and workplace locations are subjected to the analysis. The timestamps linked with the departure and arrival timings at the working location, on the other hand, are based on mobile phone usage rather than the user's real arrival and departure times. As a result, the measured times are simply a rough estimate of the true times. The gap between the departure time from the workplace on 1 day and the arrival time at the workplace the next working day is

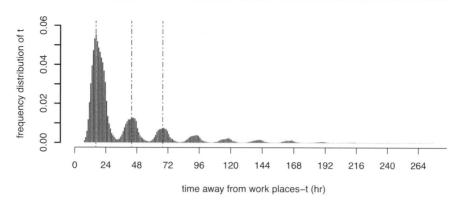

Fig. 3.4 Time away from the workplaces

regarded as the time spent away from the office. The average time spent away from work is 39.5 h, with 16.7, 22, and 46.7 h in the first, second, and third quartiles, respectively. 63.4% of the observations are focused on the first peak, as can be observed (people arriving at the workplace after being away for 32 h or less). The average duration of the first peak is 16 h, which makes sense for most people who work 8 h a day and are away for the remaining 16 h. However, only 11% of this data is collected on Monday, with the remaining observations dispersed evenly over the subsequent working days (Tuesday to Friday). The second peak accounts for 15.8% of all observations (people returning to work after being gone for 32–57 h), with Monday accounting for 31% of the data. The third peak contains 10% of the total number of observations (people arriving at their workplace after being away for 58–81 h). Surprisingly, 46% of data from the third peak is collected on Monday, with an average time of 68 h. The third surge could be attributed to those who have been gone from work since Friday evening due to the weekend.

3.2.4 Trip Detection

We grouped the sequential traces connected with the stay locations along the date and time of day for each user. If the user has more than one stay location inside a 24-h (1-day) period, a trip can be detected if midnight is used as the transition time from 1 day to the next. A trip is assumed to be made between two consecutive stay sites. As a result, the first interaction time at site i and the latest interaction time at location $i + 1$ must occur within 1 day. We looked at the number of journeys taken by 319,508 people over the course of 2 weeks, which included 10 weekdays and 4 weekends. The distribution of trips per user is shown in Fig. 3.5.

The first, second, and third quartiles of the total number of trips per user are 3, 7, and 15 trips, respectively, with corresponding first, second, and third quartiles of 2, 5, and 8 active trip making days. A day on which the user has made at least one trip is defined as an active trip making day. We divided each user's total journeys by the

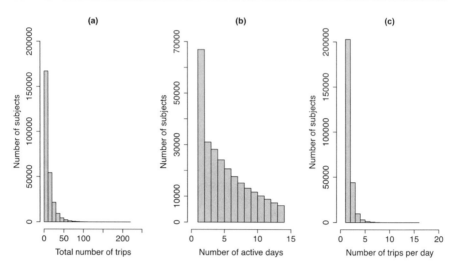

Fig. 3.5 Trips by sample users over the period of 2-weeks. (**a**) Total number of trips per user. (**b**) Total number of active trip making days per user. (**c**) Average number of trips per user per day

total number of active trip-making days to calculate the average number of trips per user per day. There were 1.3, 1.7, and 2 average trips per day in the first, second, and third quartiles, respectively.

3.2.5   Trip Types

Determining the origin and destination locations of a trip is one of the most difficult aspects of the transportation planning process. To obtain logical trip estimates, transportation demand modeling necessitates data aggregation in traffic analysis zones and correct demarcation of these zones. TAZs are largely defined in principle using criteria that propose locations with similar land use activity characteristics [2]. We were given the option of selecting one of three administrative divisions: 14 regions, 45 departments, or 123 arrondissements (districts). TAZs are defined in this study based on the smallest administrative divisions (districts), which also include demographic data (e.g., census). Senegal's national boundary serves as an exterior barrier. Trips outside the external perimeter, in other words, are not taken into account in our analysis. Trips that do not have both ends of the route inside the exterior cordon are included in this category. The 123 districts are depicted in Fig. 3.6b.

The sampled user's mobile phone records—whose traces are displayed in Fig. 3.2a are used to depict different types of trips. Inter-district and intra-district journeys are defined as trips with origins and destinations both within the study territory limited by the exterior cordon. When a user's stay position is detected at a cell tower in one district and then another stay location is detected at a cell tower in a

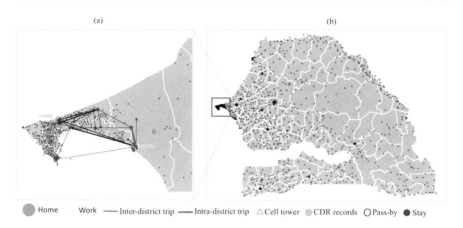

Fig. 3.6 Trip types of a sampled user (**a**) and study region (**b**)

different district, this is known as an inter-district trip (Fig. 3.6a, black lines). When the user's stay locations are detected at separate cell towers in the same district, this is referred to as an intra-district journey (Fig. 3.6a, orange lines).

Trips can be classified as home-based work (HBW), home-based other (HBO), or non-home based (NHB) depending on the type of origin and destination sites. HBW journeys are conducted between a person's home and their place of employment. The HBO trips are conducted between a person's home and other locations that are not related to work. Regardless of the objective of the trip, NHB tours do not begin or end at a person's home. The sampled user's home and work locations are determined at cell towers inside the district 1 and district 10 boundaries, respectively, using the approach described in our previous sections. All of the sampled user's other visited locations are categorized as "other". The trip to these destinations is then linked to work, home, or other activity kinds. For example, trips between the sampled user's home (Fig. 3.6a, solid orange circle) and work (Fig. 3.6a, solid yellow circle) are classified as HBW (Fig. 3.6b).

Figure 3.7a depicts inter-district trips based on 319,508 sample users over a 2-week period. The majority of inter-district journeys are performed between districts in the Dakar, Thies, and Diourbel regions (area marked in rectangle). These regions are home to 47.3% of Senegal's population and are economically, socially, and politically significant. The number of intra-district journeys is represented by a heat map in Fig. 3.7b. Large intra-district excursions have been observed in small districts within the Dakar region, which have a high population density and a large capacity for trip generation and attractiveness (31.2% of the total intra-district trips in Senegal).

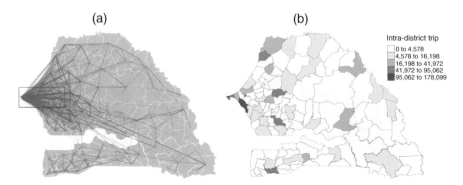

Fig. 3.7 Inter-district trips (**a**). Intra-district trips (**b**)

3.2.6   Trip Correction

According to the previous data, the bulk of users have a minimal number of daily trips. This can be explained by the fact that certain users may not use their cell phone during their travel, making it impossible to infer those trips. This could result in an inaccurate portrayal of the users' mobility behavior. We solve this problem by devising a mechanism for including portions of commuting trips that were previously missed by the OD estimate algorithm. Non-commuting trips that were not detected are not included in our analysis. Commuting trips are frequently made between the user's home and his or her workplace. We began our investigation by assuming commuting trips. We chose 243,928 people out of 319,508 who have clearly specified home and work locations.

We calculated the commute distance as the Euclidean distance between the home and work sites once the home and work locations were established. Individuals looking for work can now search for opportunities across a larger geographical area because of advancements in transportation, technology, and infrastructure [54]. Long-distance commutes may be difficult in Senegal due to low car ownership, underdeveloped public transportation, and poor urban road infrastructure.

We set a limit on the longest commute distance due to Senegal's level of developments. Mishra et al. [55] utilized a similar approach to estimate commute trips in order to mitigate the impact of long but occasional travels. ULTRANS [56] also investigated several values for limiting distance disutility for business and commuting trips to guarantee that the established model does not provide values that favor more distant locations. For some traffic engineering applications, utilizing percentiles as a threshold value is typical. A distance cap of 25 km was employed in this study, which is similar to the 85th percentile commuting distance. We additionally filter out commuting distances of less than 0.5 km, which is designed to reduce noise caused by cell towers randomly switching, particularly near tower boundaries. Kung et al. [57] employed a similar approach. While this filtering reduces the number of commuters, we concentrate on modifying the undiscovered commute trips for users who have consistent home and work locations and a manageable commute distance.

Previous research assumed that the user begins and concludes each day's journey at home. We relax this assumption in our research to account for the potential that the first/last commute trip's origin/destination could be somewhere else. In the absence of 'other-to-work' and 'work-to-other' commute trips, commute trip adjustments for undetected homebound and workbound trips are made. For each commuter, we added the following trips during the 10 working days in our dataset: (i) for a commute trip with a work location as destination but no corresponding return trip (work-to-home or work-to-other), we add a home bound trip (work-to-home); and (ii) for a commute trip with a work location as origin but no corresponding work bound trip (home-to-work or other-to-work), we add a work bound trip (home-to-work).

3.2.7 Trip Expansion

Following the estimation of the sample OD flows, the next step is to enlarge them to represent the mobility behavior of the entire population. This is accomplished through the employment of a population expansion factor, which connects sample users to the population from which the sample was obtained. We use census data from around the same time period as the CDR data as a secondary source of demographic data that may be linked to the sample. Then, for each district, the expansion factor is calculated as the ratio of the total population of that district to the number of sample users recognized as inhabitants of that district based on CDR data. A previous study by Alexander et al. [30] took a similar method, however we did not include the population age group 0–4 in our analysis.

3.2.8 Trip Distribution Modeling

Trip distribution modeling is the second component of the traditional four-step transportation forecasting approach when it comes to forecasting travel demand.

3.2.8.1 Gravity Models

A gravity model for trip distribution assumes that the trips produced at an origin and attracted to a destination are directly proportional to the total trip productions at the origin zone and the total attractions at the destination zone, and inversely proportional to the travel cost between the zones [2]. A simplest version of the gravity model for the flow between two traffic analysis zone takes the following functional form in Eq. (3.3).

$$T_{ij} = \frac{\alpha P_i P_j}{d_{ij}^2}, \tag{3.3}$$

where P_i and P_j are the populations of zone i and zone j, d_{ij} is the distance zone i and zone j, α is the gravity constant for trip distribution, and T_{ij} is the number of undirected trips between the two zones.

The model was further modified by assuming that distance is the disutility of a journey, which can be assessed by a cost function that can be a function of distance, travel time, or generalized cost between zones [2]. The model was also improved by using total trip ends (O_i and D_j) instead of total population. A doubly constrained entropy maximizing model can be written as Eq. (3.4).

$$T_{ij} = A_i O_i B_j D_j f\left(C_{ij}\right), \tag{3.4}$$

where, the single gravity constant for trip distribution factor α is replaced by two sets of balancing factors $A_i = 1/\sum_j B_j D_j f\left(C_{ij}\right)$ and $B_j = 1/\sum_i A_i O_i f\left(C_{ij}\right)$ which ensure that the estimates of T_{ij}, when summed across both rows and columns of the matrix equal the known O_i and D_j totals; $f(C_{ij})$ is a generalized cost function. Ortuzar and Willumsen [2] highlights the popular versions for cost function, $f(C_{ij})$, such as negative exponential function ($e^{-\beta C_{ij}}$), inverse power function (C_{ij}^{-n}) and combined function ($C_{ij}^{-n} e^{-\beta C_{ij}}$). Where, β and n are the exponential and power parameters for cost function, respectively.

3.2.8.2 Log-Linear Models

Log-linear models have been used to analyze the values found in contingency tables. In a similar way to Fig. 3.8, we examine the origin-destination flow between 123 districts. Figure 3.8 can be viewed as a two-way contingency table, with values in the interior cells (intra-district flows, i and inter-district flows, I) determined by the column marginal D_j, which represents each zone's trip attraction, and the row

	$DEST_1$	$DEST_2$	$DEST_3$	.	.	.	$DEST_{121}$	$DEST_{122}$	$DEST_{123}$	O_i
$ORIG_1$	i	I	I	I	I	I	I	I	I	O_1
$ORIG_2$	I	i	I	I	I	I	I	I	I	O_2
$ORIG_3$	I	I	i	I	I	I	I	I	I	O_3
.	I	I	I	i	I	I	I	I	I	.
.	I	I	I	I	i	I	I	I	I	.
.	I	I	I	I	I	i	I	I	I	.
$ORIG_{121}$	I	I	I	I	I	I	i	I	I	O_{121}
$ORIG_{122}$	I	I	I	I	I	I	I	i	I	O_{122}
$ORIG_{123}$	I	I	I	I	I	I	I	I	i	O_{123}
D_j	D_1	D_2	D_3	.	.	.	D_{121}	D_{122}	D_{123}	$\sum_i \sum_j O_i D_j$

Fig. 3.8 Sample origin-destination flow

marginal O_i, which represents each zone's trip generation. The marginal is determined by the overall volume of flows $(\sum_i \sum_j O_i D_j)$.

Flowerdew and Lovett [58] and Dennett [59] already demonstrated that the doubly constrained, multiplicative model in (4) is comparable to a statistical (additive) log-linear model. In a multiplicative spatial interaction model, Dennett [59] underlined the significance of using a calibrated statistical model to determine the distance decay parameter. The multiplicative version of an additive log-linear model describing the full system can be written as Eq. (3.5).

$$T_{ij} = \tau \tau_i^O \tau_j^D \tau_{ij}^{OD}, \tag{3.5}$$

where τ is the overall main effect τ_i^O and τ_j^D are the row and column 'main effects' represented by categorical variables with i and j are both 123 levels (total number of districts in Senegal), τ_{ij}^{OD} is the interaction component with $i = 1, \ldots, n * j = 1, \ldots,$ n parameters, where, $n = 123$ (i.e., 15,129 in total for our case). By taking the natural logarithm, the multiplicative component model in Eq. (3.5) can be expressed as a log-linear (additive) model as follows (Note that $\ln(\tau) = \lambda$).

$$ln\left(T_{ij}\right) = \lambda + \lambda_i^O + \lambda_j^D + \lambda_{ij}^{OD} \tag{3.6}$$

3.3   Results and Discussion

We examine various log-linear and gravity models for treating total flow (inter and intra-district flows), as well as trip distances computed in various ways. In all situations, we employ the method described in the Sect. 3.2.7 (trips expansion) to examine the average daily OD flow derived from sample users and enlarged to the general population.

3.3.1   Travel Distances

We estimate intra-district and inter-district trip distances using two alternative methodologies. The approach 1 uses Eq. (3.1) to calculate the intra-district journey distance. As a result, we assume that the districts' areas are circular, and the radius is calculated using the equation (*area of district = πr^2*). The matched inter-district journey distance is calculated using the Euclidean distance between the origin and destination districts' centroids. In the approach 2, we adopt Eq. (3.2) for both intra-district and inter-district journey distances.

Figure 3.9a and b show the average daily intra-district and inter-district trip distances measured through the first and second approaches, respectively. Figure 3.9a exhibits a longer intra-district trip distance compared to the intra-district distance in Fig. 3.9b. The intra-district trip distance in Fig. 3.9a is estimated as a

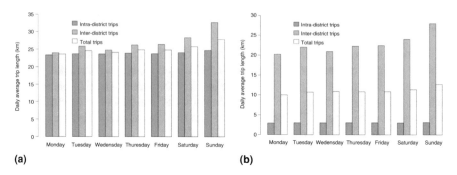

Fig. 3.9 Trip distances. Approach 1 (**a**). Approach 2 (**b**)

Table 3.1 List of trip distribution models

Model	Model type	Inter-district trip distance	Intra-district trip distance
Model 1	Log-linear	Approach 1	NA
Model 2	Log-linear	Approach 2	NA
Model 3	Gravity	Approach 1	NA
Model 4	Gravity	Approach 2	NA
Model 5	Log-linear	Approach 1	Approach 1
Model 6	Log-linear	Approach 2	Approach 2
Model 7	Gravity	Approach 1	Approach 1
Model 8	Gravity	Approach 2	Approach 2

function of the area of a district instead of the intrinsic features that influence trip such as time of day, trip purpose, and other district attributes [4]. As a result, the average daily intra-district and inter-district journey distances have become unrealistically identical. A similar pattern of considerable trip distance differences between intra-zonal and inter-zonal journeys was discovered in a previous study [6].

3.3.2 Trip Distribution Models

Table 3.1 displays the results of eight different trip distribution models. Intra-district journeys are not taken into account in the first series of models (Models 1–4). Intra-district travels are originally allocated to neighboring districts to equalize the origin and destination trip totals. The distribution is made based on the propensity of inter-district trips from district i to the remaining 122 districts using $(T_{ij} \times O_{ii})/O_{ib}$, where, O_{ii} is intra-district trips of district i, T_{ij} is the number of trips from district i to district j, and O_{ib} is origin total of district i before the addition of O_{ii}. No intra-district trip distances are applicable in the first four models as the intra-district trips are not modeled explicitly. The second group of models, Model 5 to Model 8, focus on total

OD flow (both the inter-district and intra-district OD flows), where intra-district trips are modeled explicitly.

A total of 15,129 inter-district and intra-district OD flows were generated using a 123-district level zoning system. Our research shows that there are no empty zones with no trip productions or attractions. Our data also contains no null cells, which represent unobservable movements. We suppose that our observation is based on the presence of feasible connections connecting all of the zones. Our dataset, on the other hand, contains a substantial number of zero cells. These cells are handled as matrix cells with reported motions but no observation. The zero cells are taken into account in our calculations. Zero observations are treated as a likely consequence of a modest but finite trip probability in the log-linear modeling approach [60].

Starting with Model 1's estimation, categorical variables related with trip origins and destinations, as well as other explanatory factors important to the origin-destination interaction, are used to fit a doubly constrained log-linear model. We used the inverse power and negative exponential functions to calculate the travel cost function. The findings of the inverse power function are discussed because it produced a better fit. If we recall Eq. (3.5) and replace the interaction component with the inverse power function, we obtain a new model with the form of Eqs. (3.7) or (3.8).

$$T_{ij} = \tau \tau_i^O \tau_j^D C_{ij}^{-n} \tag{3.7}$$

$$\ln\left(T_{ij}\right) = \lambda + \lambda_i^O + \lambda_j^D - n\log\left(C_{ij}\right) \tag{3.8}$$

An exploratory analysis on our data suggested the presence of overdistribution. Further analysis is carried out to check for distribution using Cameron and Trivedi methodology [61], with the results confirming the presence of overdistribution. In order to deal with overditribution, we adopt a quasi-poisson model in a generalized linear model (GLM) framework. To fit the log-linear model in Eq. (3.8), we use the *glm()* function in R [62]. A total of 15,006 inter-district flows were created using a 123-district level zoning system. We also use a total of 122 categorical variables, 122 for the origin and another 122 for the destination (one of the variables from each group is used as a reference category). The travel cost is also represented by a continuous (non-categorical) trip distance variable. In the GLM framework, the basic poisson model posits that the predicted value of OD trips follows a Poisson distribution with a mean that is logarithmically related to a linear combination of the origin and destination category variables, as well as the distance variable. Further-more, because the variance in the poisson model is the same as the mean, the dispersion parameter is set to 1. To deal with overdispersion, the dispersion parameter in the quasi-poisson model is unrestricted. Further explanation of how to model over-dispersed count data and the structure of the accompanying models can be found in Zeileis et al. [63] and Hoef et al. [64].

Model 1's estimation is shown in Eq. (3.9), where $\widehat{T_{ij}}$ is the estimated flow between districts i and j. The origin-specific (*orig*) and destination-specific (*dest*) variables were compared to the reference categories $orig_1$ and $dest_1$, respectively, to determine their relative relevance. The origin-specific categorical variables' coefficients range from 4.99 (Koussanar), which is one of the districts with a high number of daily trips, to -0.94 (Rufisque), which is one of the districts with a low number of daily travels. The destination-specific categorical variables have coefficients ranging from 0.75 (Dakar Plateau) to -4.18 (Karantaba), one of the districts with few daily travels. The distance variable's coefficient is negative and substantial. The effect of distance is relatively predictable, given that the interaction between origin and destination districts decreases as the distance between them grows due to rising trip costs.

$$\widehat{T_{ij}} = exp \left(14.20 + 1.44orig_2 + 0.21orig_3 + 1.25orig_4 + \ldots + 4.62orig_{123} \right.$$
$$\left. + 0.17dest_2 - 0.10dest_3 + 0.75dest_4 + \ldots - 1.04dest_{123} - 2.40 \, ln \, C_{ij} \right)$$

$$(3.9)$$

Equation (3.10) depicts Model 2 estimation. The model parameters and coefficients are analyzed in the same way as Model 1.

$$\widehat{T_{ij}} = exp \left(13.05 + 1.25orig_2 + 0.30orig_3 + 1.68orig_4 + \ldots + 3.14orig_{123} \right.$$
$$\left. + 0.05dest_2 - 0.07dest_3 + 1.08dest_4 + \ldots - 1.58dest_{123} - 1.98 \, ln \, C_{ij} \right)$$

$$(3.10)$$

Estimation of Model 3, a doubly constrained gravity model for inter-district OD flows, necessitates estimation of its parameters such that a reasonable benchmark trip pattern can be reproduced. We utilize the model in Eq. (3.4), which has the parameters A_i, B_j, and a parameter for the trip distance-based travel cost. There are 122 A_i parameters, 122 B_j parameters, and one travel cost parameter in general (in total there are 123 districts i.e., zones). The parameters A_i and B_j are calibrated during the gravity model calculation, whereas the journey cost (trip distance) parameter is obtained from the Eq. (3.9).

For the models that do not account for intra-district trips, Fig. 3.10 illustrates a comparison of observed and estimated inter-district trips.

The results of the log-linear models described in Eqs. (3.9) and (3.10), respectively, are shown in Fig. 3.10a and b. The findings of the related gravity models to the log-linear models in (3.9) and (3.10), respectively, are shown in Fig. 3.10c and d. We also calculated a more traditional R-Squared (R^2) value to compare observed visits to model outputs, and obtained 0.763, 0.883, 0.763, and 0.883 for Models 1, 2, 3, and 4, respectively. The effect of travel cost on the estimated and observed visits is also shown in Fig. 3.10. When the trip distance is less, there is more interaction between the origin and destination areas.

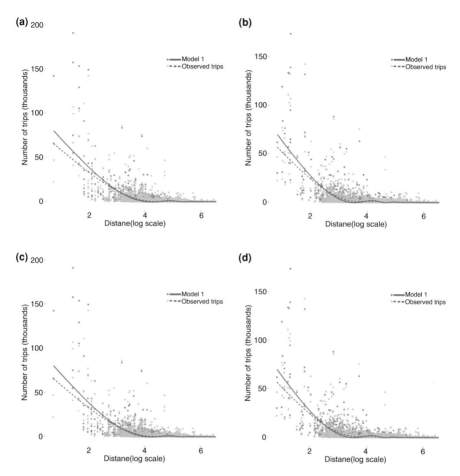

Fig. 3.10 Observed vs estimated inter-district trips. (**a**) Model 1. (**b**) Model 2. (**c**) Model 3. (**d**) Model 4

Figure 3.11 depicts a comparison of observed and estimated trips for models in which intra-district travels are explicitly modeled. The findings of the log-linear models described in (11) and (12) are shown in Fig. 3.11a and b. The findings of the relevant gravity models (Model 7 and Model 8) are shown in Fig. 3.11c and d.

$$\widehat{T_{ij}} = exp\left(14.34 + 1.67orig_2 + 0.46orig_3 + 0.99orig_4 + \ldots + 4.96orig_{123}\right.$$

$$\left. + 0.88dest_2 + 0.18dest_3 + 0.78dest_4 + \ldots - 1.12dest_{123} - 2.72\ln C_{ij}\right)$$

$$(3.11)$$

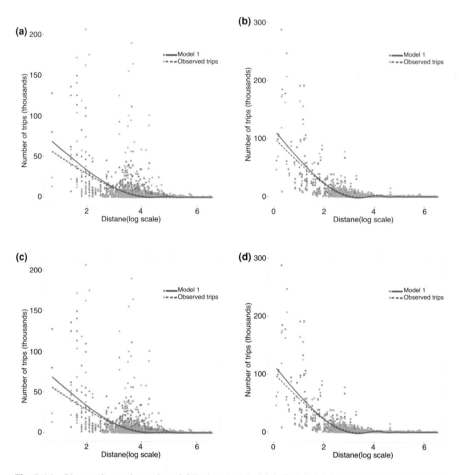

Fig. 3.11 Observed vs estimated total OF trips. (**a**) Model 5. (**b**) Model 6. (**c**) Model 7. (**d**) Model 8

We also computed the R-Squared (R^2) value to compare the observed trips against the model outputs and we found 0.43, 0.94, 0.43, and 0.94 for Model 5, Model 6, Model 7, Model 8, respectively.

$$\widehat{T_{ij}} = exp\left(11.79 + 1.56 orig_2 + 0.72 orig_3 + 0.71 orig_4 + \ldots + 3.35 orig_{123}\right.$$

$$\left. + 0.03 dest_2 + 0.12 dest_3 + 0.64 dest_4 + \ldots - 0.85 dest_{123} - 1.88 \ln C_{ij}\right)$$

$$(3.12)$$

We estimated eight distinct models. In the case of Models 6 and 8 ($R^2 = 0.94$), a better fit between observed and estimated OD flows is obtained. The intra-district and inter-district journey distances were calculated using (2). Model 5 and Model 7 produced a poor fit with $R^2 = 0.43$.

3.3.3 Log-Linear Model-Based Approaches

The trip distance or travel cost (lnC_{ij}) parameters of the four doubly constrained log-linear models are reported in Table 3.2, and the section on trip distances describes the two alternative ways to estimating the intra-district and matching inter-district trip distances. All of the models have a comparable negative component matching to the trip distance parameters, indicating that as the trip distance grows, the level of origin/destination interaction decreases. However, between the models calculated for inter-district OD flow (Model 1 and Model 2), as well as inter-district and intra-district OD flow (Model 5 and Model 6), there is a variation in the value of the trip distance parameter.

Model 1 has a trip distance parameter of -2.397 that does not account for intra-district trips. Model 5, which has a similar trip distance specification to Model 1, but which explicitly models intra-district trips, has a smaller trip distance parameter (-2.716). The explicit reference of intra-district journeys, on the other hand, has improved the trip distance parameter from -1.983 (Model 2) to -1.884 (Model 6). In Model 5, the average intra-district journey distance is calculated using Eq. (3.1), which is a function of a district's area. As a result, regardless of the underlying characteristics that influence journeys, intra-district trips have a long trip distance as long as the district is vast (Fig. 3.9a). As a result, the model is overly sensitive to the distance-decay effect, resulting in an overestimation of people's disutility toward lengthier trips. Model 6 calculates the length of an intra-district journey using Eq. (3.2), which is consistent with the reality that intra-district trips are substantially shorter than inter-district trips (Fig. 1.9b). Because intra-district travels are explicitly modeled, the value of the trip distance parameter increases from -1.983 to -1.884. When compared to Model 6, Model 5 generates fewer trips due to its reduced trip distance parameter. This suggests that journey distance influences decision-making, and people may prefer to make shorter trips inside their district.

Table 3.2 Result summary of log-linear models

Model	Trip-distance estimation	Trip categories	Log-linear model			
			Intercept	lnC_{ij}	Deviance reduction	R^2
Model 1	Approach 1	Inter-district	14.195	−2.387	87.70%	0.763
Model 2	Approach 2	Inter-district	13.053	−1.977	91.70%	0.883
Model 5	Approach 1	Inter and intra districts	14.339	−2.716	80.00%	0.430
Model 6	Approach 2	Inter and intra districts	11.794	−1.884	94.80%	0.940

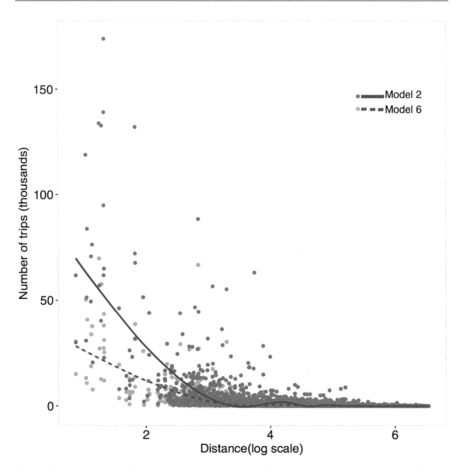

Fig. 3.12 Comparison of the number of estimated inter-district trips by two sets of trip distribution models

The deviation statistics are also used to evaluate the log-linear models' goodness-of-fit. Null deviance and residual deviance are shown in Table 3.2. The worst-performing model (Model 5) is responsible for 80% of the null deviation, while the best-performing model (Model 6) is responsible for 94.8%.

To show the difference between the two groups of trip distribution models, we compared estimated inter-district trips based on Model 2 and Model 6 (Model 2—does not account for intra-district trips; Model 6—intra-district trips are modeled explicitly).

Both models in Fig. 3.12 exhibit a similar pattern of a high number of inter-district trips interactions between districts when the trip distance is short, and the interaction disappears as the trip distance increases. Model 2, on the other hand, provides more inter-district travel. The exclusion of intra-district travels is one of the reasons. As a result of the disproportionate distribution of trips to neighboring districts, more trips than necessary are directed across the inter-district network. The findings imply that explicitly including intra-district travels in trip distribution models increases the accuracy of the outputs.

3.3.4   Trip Distance Distribution

For all of the models, we compared the observed and estimated journey distances. The distributions of travel distances are shown in Fig. 3.13.

The similarities between the estimated and observed trip distance distributions in Fig. 3.13d in terms of reproducing the observed inter-district and intra-district OD flows, gives a good indicator of the accuracy of the log-linear and gravity trip distribution models (Model 6 and Model 8). We use the coincidence ratio to quantify

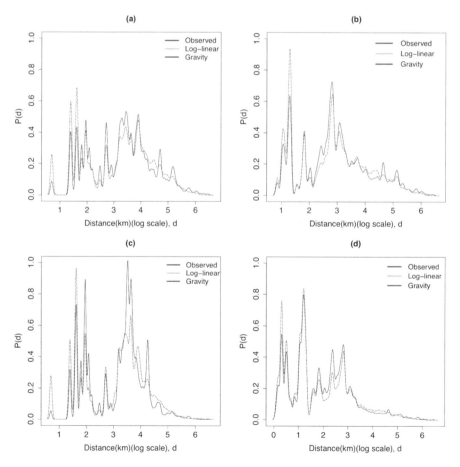

Fig. 3.13 Trip distance distribution of observed and estimated OD flows. (**a**) Model 1 and Model 3. (**b**) Model 2 and Model 4. (**c**) Model 5 and Model 7. (**d**) Model 6 and Model 8

Table 3.3 Coincidence ratio values (d denotes trip distance in log scale)

Model	$d \leq 1$	$1 < d \leq 2$	$2 < d \leq 3$	$3 < d \leq 4$	$d > 4$	Average
Model 1	0.307	0.771	0.756	0.882	0.957	0.735
Mode 2	0.711	0.764	0.813	0.963	0.987	0.847
Model 3	0.307	0.770	0.756	0.883	0.958	0.735
Model 4	0.711	0.764	0.813	0.963	0.986	0.847
Model 5	0.171	0.897	0.850	0.888	0.890	0.739
Model 6	0.782	0.991	0.777	0.960	0.969	0.896
Model 7	0.171	0.897	0.851	0.888	0.891	0.739
Model 8	0.781	0.991	0.777	0.960	0.973	0.897

how well the estimated distribution curve coincides with the observed one, in addition to visual comparison of the estimated and observed trip distances. The coincidence ratio is defined as a number between 0 and 1, with values closer to 1 suggesting that the model is performing reasonably [55]. The coincidence ratio can be calculated using the following formula.

$$CR = \frac{\sum_t \min (obs_t, est_t)}{\sum_t \max (obs_t, est_t)}, \tag{3.13}$$

where CR represents the coincidence ratio, obs_t represents the proportion of observed distribution in interval t, and est_t represents the proportion of estimated distribution in interval t. For five trip distance intervals, the CR is determined (log scale). The calculated models perform well, as shown in Table 3.3 with an average coincidence ratio of 0.73. However, in the short trip distance period ($d \leq 1$), four of the model estimations based on trip distance assessed using the approach 1 did not perform well.

3.4   Conclusion

Increasing mobility demand has put strain on existing mobility infrastructure, such as roads and public transportation services, resulting in traffic congestion and shaky public transportation systems. To mitigate these effects, effective planning and operation are essential as the mobility needs continue to expand. Because of the restricted budget available to conduct large-scale mobility surveys in developing countries, obtaining mobility data is a difficult undertaking [65]. In this chapter, we evaluate country-wide people migration patterns using CDR data. The primary goal of this study is to investigate the potential of CDR data for detecting the origin and destination of trips, as well as to create trip generation models and assess the impact of intra-zonal travels on model estimates.

In this chapter, we estimated eight different models, which can be divided into categories based on the type of travel patterns (do not account for intra-district trips or explicitly model intra-district trips); modeling approach (gravity or log-linear); or how the travel cost is formulated (trip distance measured between districts centroids or between cellular network tower locations). Since assessing intra-zonal trips and the corresponding travel cost are challenging, intrazonal trips are typically not included in trip distribution modeling. By analyzing Senegal's CDR data at the 123-district level traffic analysis zone system, we discussed two methods for estimating intra-zonal travel distance. When the average intra-zonal journey distance was calculated as a function of the zone's area, the trip distances were found to be considerably longer. As a result, the estimated model is extremely susceptible to distance degradation. In the second method, we divided each of the 123 zones into subzones in order to calculate intra-zonal journeys and costs. The average intra-zonal journey and travel cost were calculated using the positions of cellular network towers within each zone as the subzone centers. The average intra-zonal journey is much shorter than the inter-zonal excursions when using this method. As a result, the estimated model is more resistant to the distance decay effect.

This analysis is useful to transport planners in developing nations since it explains how mobile phone communication data may be used to deduce a country's mobility patterns. The findings can also be utilized to inform judgments on inter-district public transportation planning and large national and regional road network expansion projects. Planners in developed countries can benefit from this research study in two ways: (i) the described modeling approach can complement regional transportation travel demand models by incorporating mobility data from under-surveyed locations in a region; and (ii) it provides a foundation for using mobile phone data to continuously update traditional urban data, which is usually expensive to collect. Furthermore, the study's findings offer a supplemental and/or alternative method of acquiring intra-zonal trip data and estimating processes. This is especially relevant given the growing global interest in making cities more walkable and bikeable, since a large percentage of trips are intra-zonal. Most current techniques of gathering intrazonal trip information, which are based on pedestrian counts, either manually or automatically, are not cost-effective [66, 67]. Furthermore, this technique can overcome the constraints of fixed zoning regimes that do not account for continuing land use changes (spatially and temporally). Trips inferred from CDR data are only linked to their respective tower sites, not to any zoning scheme. Future travel demand studies using CDR data could result in the development of a new zoning scheme that better suits the study scope and goals, giving transportation analysts more freedom [2].

Despite the study's significant benefits, there are a few limitations to be mentioned. We tried to associate CDR data to either an origin/destination zone or a transient zone travelled through on the way to an origin/destination zone. This was calculated depending on how long a person stayed in a specific location (cell tower). The 10 min threshold value used to separate stay and pass-by areas was chosen arbitrarily. Future research should look towards creating a dynamic threshold number that takes into account the extent of cell tower coverage, traffic, and pedestrian realities in a given region. Future research using CDR data with triangulated coordinates may leverage the stay detection algorithms developed by Zheng et al. [14] and Hariharan and Toyama [15]. The expansion factor necessary to spread the results to a general population is calculated using the residence of sample users in our analysis. Future studies, on the other hand, could include complete profiles of sample users, such as socioeconomic and demographic data, in order to accurately depict the makeup of sample data. One of the goals of this research is to give transportation planners, particularly in developing countries, an option to consider in the absence of gathered transportation data when estimating trip demand for transportation system design and planning. However, due to the lack of large-scale mobility surveys in Senegal, the given data cannot be compared to ground truth. Gathering and using a ground truth data to validate projected OD flows and the effect of intra-district travels on trip distribution modeling could be considered for a future study. There were no notable holidays or national events between January 7, 2013 and January 20, 2013, the dates of our analysis. Our assumptions and evaluations, on the other hand, are based on the 2-week CDR data's duration and spatial resolution. With CDR data over a longer time, the representativeness of the extracted mobilities and flows could be improved.

References

1. Johnson JH. Inter-urban migration in Britain: a geographical perspective. London: Taylor & Francis; 1984.
2. de Dios Ortúzar J, Willumsen LG. Modelling transport. Chichester, UK: Wiley; 2011.
3. Cascetta E, Pagliara F, Papola A. Alternative approaches to trip distribution modelling: a retrospective review and suggestions for combining different approaches. Pap Reg Sci. 2007;86(4):597–620.
4. Becker RA, et al. A tale of one city: using cellular network data for urban planning. IEEE Pervasive Comput. 2011;10(4):18–26.
5. Bhatta BP, Larsen OI. Are intrazonal trips ignorable? Transp Policy. 2011;18(1):13–22.
6. Greenwald MJ. The relationship between land use and intrazonal trip making behaviors: evidence and implications. Transp Res Part D Transp Environ. 2006;11(6):432–46.
7. Nair HS, Bhat CR. Modeling trip duration for mobile source emissions forecasting. J Transp Stat. 2003;6(1):17–32.
8. Venigalla MM, Chatterjee A, Bronzini MS. A specialized equilibrium assignment algorithm for air quality modeling. Transp Res Part D Transp Environ. 1999;4(1):29–44.
9. Cervero R, Duncan M. Walking, bicycling, and urban landscapes: evidence from the San Francisco bay area. Am J Public Health. 2003;93(9):1478–83.
10. Qin X, Ivan JN. Estimating pedestrian exposure prediction model in rural areas. Transp Res Rec J Transp Res Board. 2001;1773(1):89–96.
11. Miranda-Moreno LF, Morency P, El-Geneidy A. How does the built environment influence pedestrian activity and pedestrian collisions at intersections. In: TRB (Transportation Research Board), the 89th annual meeting of the Transportation Research Board. 250, 8. Washignton DC; 2010. p. 10–14.
12. Boarnet MG, Anderson CL, Day K, McMillan T, Alfonzo M. Evaluation of the California safe routes to school legislation. Am J Prev Med. 2005;28(2):134–40.
13. Demissie MG. Combining datasets from multiple sources for urban and transportation planning: emphasis on cellular network data. Coimbra: Coimbra Univ.; 2014.
14. Zheng Y, Zhang L, Xie X, Ma W-Y. Mining interesting locations and travel sequences from GPS trajectories. In: Proceedings of the 18th international conference on world wide web (WWW '09); 2009. p. 791–800.

15. Hariharan R, Toyama K. Project lachesis: parsing and modeling location histories. Geogr Informat Sci. 2004;3234:106–24.
16. Liao L, Patterson DJ, Fox D, Kautz H. Learning and inferring transportation routines. Artif Intell. 2007;171(5–6):311–31.
17. Combe C, Grandet R, Igoudjil I, Paderi R. The impact of urban data on daily mobility. 2014.
18. Vaccari A, et al. A holistic framework for the study of urban traces and the profiling of urban processes and dynamics. In: 2009 12th international IEEE conference on intelligent transportation systems; 2009. p. 1–6.
19. Caceres N, Wideberg JP, Benitez FG. Review of traffic data estimations extracted from cellular networks. IET Intell Transp Syst. 2008;2(3):179–92.
20. Calabrese F, Colonna M, Lovisolo P, Parata D, Ratti C. Real-time urban monitoring using cell phones: a case study in Rome. IEEE Trans Intell Transp Syst. 2011.
21. Ratti C, Sevtsuk A, Huang S, Pailer R. Mobile landscapes: graz in real time. In: 3rd symposium on LBS and telecartography; 2005. p. 28–30.
22. Bar-Gera H. Evaluation of a cellular phone-based system for measurements of traffic speeds and travel times: a case study from Israel. Transp Res Part C Emerg Technol. 2007.
23. Demissie MG, de Almeida Correia GH, Bento C. Intelligent road traffic status detection system through cellular networks handover information: an exploratory study. Transp Res Part C Emerg Technol. 2013.
24. Liu HX, Danczyk A, Brewer R, Starr R. Evaluation of cell phone traffic data in Minnesota. Transp Res Rec J Transp Res Board. 2008;2086(1):1–7.
25. Demissie MG, Antunes F, Bento C, Phithakkitnukoon S, Sukhvibul T. Inferring origin-destination flows using mobile phone data: a case study of Senegal. In: 2016 13th international conference on electrical engineering/electronics, computer, telecommunications and information technology (ECTI-CON); 2016. p. 1–6.
26. Calabrese F, Di Lorenzo G, Liu L, Ratti C. Estimating origin-destination flows using mobile phone location data. IEEE Pervasive Comput. 2011.
27. White J, Wells I. Extracting origin destination information from mobile phone data. In: Road transport information and control, vol. 486; 2002. p. 30–4.
28. Demissie MG, Correia G, Bento C. Analysis of the pattern and intensity of urban activities through aggregate cellphone usage. Transp A Transp Sci. 2015.
29. Toole JL, Ulm M, González MC, Bauer D. Inferring land use from mobile phone activity. In: Proceedings of the ACM SIGKDD international conference on knowledge discovery and data mining; 2012. p. 1–8.
30. Alexander L, Jiang S, Murga M, González MC. Origin–destination trips by purpose and time of day inferred from mobile phone data. Transp Res Part C Emerg Technol. 2015;58:240–50.
31. Çolak S, Alexander LP, Alvim BG, Mehndiratta SR, González MC. Analyzing cell phone location data for urban travel. Transp Res Rec J Transp Res Board. 2015;2526(1):1–17.
32. Demissie MG, Phithakkitnukoon S, Sukhvibul T, Antunes F, Gomes R, Bento C. Inferring passenger travel demand to improve urban mobility in developing countries using cell phone data: a case study of Senegal. IEEE Trans Intell Transp Syst. 2016;17(9):2466–78.
33. Gundlegård D, Rydergren C, Breyer N, Rajna B. Travel demand estimation and network assignment based on cellular network data. Comput Commun. 2016;95:29–42.
34. Lu Y, Liu Y. Pervasive location acquisition technologies: opportunities and challenges for geospatial studies. Comput Environ Urban Syst. 2012.
35. Hoteit S, Secci S, Sobolevsky S, Ratti C, Pujolle G. Estimating human trajectories and hotspots through mobile phone data. Comput Netw. 2014;64:296–307.
36. Csáji BC, et al. Exploring the mobility of mobile phone users. Phys A Stat Mech its Appl. 2013;392(6):1459–73.
37. Jiang S, Fiore GA, Yang Y, Ferreira J, Frazzoli E, González MC. A review of urban computing for mobile phone traces. 2013.
38. Schneider CM, Belik V, Couronné T, Smoreda Z, González MC. Unravelling daily human mobility motifs. J R Soc Interface. 2013;10(84):20130246.

39. Alexander LP, González MC. Assessing the impact of real-time ridesharing on urban traffic using mobile phone data. UrbComp. 2015:1–9.
40. Jiang S, Ferreira J, Gonzalez MC. Activity-based human mobility patterns inferred from mobile phone data: a case study of Singapore. IEEE Trans Big Data. 2016;3(2):208–19.
41. Toole JL, Colak S, Sturt B, Alexander LP, Evsukoff A, González MC. The path most traveled: travel demand estimation using big data resources. Transp Res Part C Emerg Technol. 2015;58: 162–77.
42. Yan X-Y, Zhao C, Fan Y, Di Z, Wang W-X. Universal predictability of mobility patterns in cities. J R Soc Interface. 2014;11(100):20140834.
43. Yang Y, Herrera C, Eagle N, González MC. Limits of predictability in commuting flows in the absence of data for calibration. Sci Rep. 2015;4(1):5662.
44. Wang Y, Homem G, Romph E. National and regional road network optimization for Senegal using mobile phone data. In: Data for development (D4D) challenge at NetMob 2015; 2014. p. 1–27.
45. Moeckel R, Donnelly R. Simulation of intrazonal traffic flows: the end of lost trips. In: Proceedings of the 11th conference on computers in urban planning and urban management (CUPUM); 2009. p. 16–8.
46. Ghareib AH. Different travel patterns: interzonal, intrazonal, and external trips. J Transp Eng. 1996;122(1):67–75.
47. Kordi M, Kaiser C, Fotheringham AS. A possible solution for the centroid-to-centroid and intra-zonal trip length problems. In: Multidisciplinary research on geographical information in Europe and beyond; 2012. p. 147–52.
48. B. of P. Roads. Calibrating and testing a gravity model for any size urban area. Washington, D.C.; 1983.
49. Batty M. Urban modelling: algorithms, calibrations, predictions. London: Cambridge: Cambridge University Press; 1976.
50. Geohive. Senegal census data. 2014. http://www.geohive.com/cntry/senegal_ext.aspx.
51. Deloitte and GSMA. Sub-Saharan Africa mobile observatory. 2012. http://www.gsma.com/publicpolicy/wp-content/uploads/2012/03/SSA_FullReport_v6.1_clean.pdf.
52. de Montjoye Y-A, Smoreda Z, Trinquart R, Ziemlicki C, Blondel VD. D4D-Senegal: the second mobile phone data for development challenge. arXiv. 2014;1407(4885):1–11. [Online]. https://arxiv.org/pdf/1407.4885.pdf.
53. Phithakkitnukoon S, Smoreda Z, Olivier P. Socio-geography of human mobility: a study using longitudinal mobile phone data. PLoS One. 2012;7(6):e39253.
54. Yapa L, Polese M, Wolpert J. Interdependencies of commuting, migration, and job site relocation. Econ Geogr. 1971;47(1):59–72.
55. Mishra S, Wang Y, Zhu X, Moeckel R, Mahaparta S. Comparison between gravity and destination choice models for trip distribution in Maryland. Transp Res Board. 2013; (January):1–22.
56. ULTRANS, H. Specto, and HBA Specto. CSTDM09-model development: long distance personal travel model. 2011(May). CA.
57. Kung KS, Greco K, Sobolevsky S, Ratti C. Exploring universal patterns in human home-work commuting from mobile phone data. PLoS One. 2014;9(6):e96180.
58. Flowerdew R, Lovett A. Fitting constrained Poisson regression models to interurban migration flows. Geogr Anal. 1988;20(4):297–307.
59. Dennett A. Estimating flows between geographical locations: get me started in'spatial interaction modelling. London; 2012.
60. Shrewsbury J. Calibration of trip distribution by generalised linear models. Univ. Canterbury; 2012.
61. Cameron AC, Trivedi PK. Regression-based tests for overdispersion in the Poisson model. J Econ. 1990;46(3):347–64.
62. R. D. C. T. 3.5.1. A language and environment for statistical computing. R Foundation for Statistical Computing; 2018.

63. Zeileis A, Kleiber C, Jackman S. Regression models for count data in *R*. J Stat Softw. 2008;27 (8):1–25.
64. Ver Hoef JM, Boveng PL. Quasi-Poisson vs. negative binomial regression: how should we model overdispersed count data? Ecology. 2007;88(11):2766–72.
65. Williams B. Sustainable urban transport in Africa: issues and challenges. 2011. https://www.itu. int/en/ITU-%0AD/Statistics/Documents/facts/ICTFactsFigures2013-e.pdf.
66. Pulugurtha SS, Repaka SR. Assessment of models to measure pedestrian activity at signalized intersections. Transp Res Rec J Transp Res Board. 2008;2073(1):39–48.
67. Schneider RJ, Arnold LS, Ragland DR. Pilot model for estimating pedestrian intersection crossing volumes. Transp Res Rec J Transp Res Board. 2009;2140(1):13–26.

Inferring and Modeling Migration Flows Using Mobile Phone CDR Data

4

Abstract

Understanding the causes and impacts of migration, as well as implementing policies aimed at providing certain services, requires estimating migration flows and forecasting future patterns. Over time, less study has been done on modeling migration flows than has been done on modeling other types of flows, such as commutes. One of the biggest hurdles to empirical analysis and theoretical developments in the modeling of migration flows has been a lack of data. Because a migration trip is far less frequent than a commute, it necessitates a longitudinal set of data for study. The data from a large mobile phone network is used in this chapter to infer migration trips and their distribution. Intra/inter-district migration flows, migration distance distribution, and origin-destination (O-D) movements are among the interesting properties of the inferred migration trips. The log-linear model, classic gravity model, and recently developed radiation model are investigated for migration trip distribution modeling, with distinct approaches applied in setting parameters for each model. As a result, among the different models, gravity and log-linear models with a direct distance (displacement) as a trip cost and district centroids as reference points perform the best. Among the radiation models, a model that considers district population is the best performing model, but not as good as the gravity and log-linear models. This chapter reflects the idea and thinking process of our original work by Phithakkitnukoon et al. (IEEE Access. 2022;10:23248–58; IEEE international conference on privacy, security, risk and trust and IEEE international conference on social computing (PASSAT/SocialCom 2011); 2011. p. 515–20), and Hankaew et al. (IEEE Access. 2019;7(1):164746–58).

Keywords

Call detail records · Migration trip inference · Trip distribution modeling · Gravity model · Log-linear model · Radiation model

S. Phithakkitnukoon, *Urban Informatics Using Mobile Network Data*, https://doi.org/10.1007/978-981-19-6714-6_4

4.1   Motivation and State of the Art

Human movement across short and long distances, both individually and in groups, has long been studied. It is critical to comprehend and model movement in order to accurately forecast human mobility. Human beings' way of life has always been intricately linked to their movement. Earlier movements were primarily impacted by variables such as climate change and hostile landscapes, but modern movements are primarily driven by socioeconomic concerns such as employment, living conditions, and food.

Short-distance journeys are mainly routine trips [1] such as the daily commute to work and shopping, which can be done from home or not, whereas long-distance trips are mostly related to a temporary or permanent change of location. Permanent relocation, sometimes known as migration, is a sort of long-distance journey affected by sociopolitical, economic, and ecological variables [2], such as job prospects and family. Long-distance travel is undertaken for a variety of reasons, including business and pleasure, and particularly includes travel by automobile, train, and airplane. Long-distance journeys which make up a substantially smaller percentage of overall travel, are less frequent, and are underrepresented in most regional and national travel demand models. It is critical, however, that these trips be incorporated in travel demand models. Long-distance travels of more than 80 km, for example, account for only 2.3% of all trips but nearly a third of total vehicle kilometers driven in the United Kingdom [3]. Because long-distance journeys make for a significant fraction of total kilometers traveled, especially on high-capacity routes, long-distance trip infrastructure is costly and time-consuming to construct. In addition, log-distance trips have a significant and growing environmental impact [4]. As a result, proper planning and operation of these trips is essential to minimize their negative effects.

Trip distribution modeling is the second of four steps in the widely used four-step transportation planning process for forecasting travel demand, which includes trip generation, trip distribution, mode choice, and route assignment, and is designed to answer questions like how many trips will be generated?, when and where will these journeys take place?, what is the means of transportation for each trip?, and what is each trip's route decision, accordingly [5]. We look into long-distance trips in this chapter, with a special focus on migration trip distribution modeling, which aims to examine and fundamentally describe the distribution of these types of long-distance trips using statistical models. Modeling long-distance travels is more difficult than commuting trips due to their unpredictability; as a result, the amount of data collected on this type of trip is restricted.

Travel demand and other human behavioral modeling are typically drawn from real-world observational data via revealed preference surveys. In some circumstances, passengers' evaluations of transportation services are analyzed using stated preference surveys, particularly when there are hypothetical transportation service alternatives and novel qualities. Most surveys collect information about a person, their household, and a daily journal of their travels. A travel survey is often conducted using a traffic count, a roadside interview, and a questionnaire. Due to the significant expenses and time commitment, major travel surveys are normally undertaken once every 10 years [6]. Although a travel survey can provide precise mobility information, it can also be inaccurate due to inaccurate responses to travel survey questionnaires, which typically rely on recalling some information from previous journeys.

The great majority of prior studies on migration flows estimate migration flows from census and population register data. These aggregate statistics may be useful for describing broad international migration patterns, urban-rural migration patterns, and some general migration patterns [7]. Censuses and population registers, on the other hand, frequently fail to capture patterns of temporary and seasonal migratory movements, particularly in years between censuses and for recent trends [8, 9]. Censuses are biased toward documented citizens, according to Massey and Capoferro [10], and may not track migration flows connected with undocumented or international migrants. Furthermore, migration-specific surveys take time and money (due to the possibility of longitudinal interviews with migrants) for most planners and government agencies [11]. Big data sources, on the other hand, can track the movements of a huge section of the population with unparalleled spatial and temporal precision [12].

Sensors such as GPS tracking units are increasingly being employed in travel surveys as a result of recent improvements in ICT [13]. However, gathering such data on a broad scale is challenging and complicated due to privacy considerations and restrictions, such as the EU general data protection regulation (GDPR). Recent initiatives have resulted in data that is restricted to a specific type of tracked individual, such as university students [14] and consumers of a single service provider for whom the data was gathered in exchange for some incentives [15]. Because of privacy considerations, this type of precise mobility data is rarely available to be used widely, making it difficult to use for extensive trip distribution modeling.

Recently, the emphasis has switched to the use of opportunistic sensing data derived from a variety of sources to provide insights on the spatial distribution and temporal evolution of people's movements. Data acquired for one reason that also presents an opportunity for another is known as opportunistic sensing data. Call detail records (CDR) from mobile phones are a sort of opportunistic sensing data that was originally collected for billing purposes but can potentially be used for human mobility research. The communication (e.g., call duration, timestamp, callers and callees identifications) and location of the linked cellular tower are recorded for billing when a mobile phone user connects to the cellular network by making or receiving a phone call or accessing the internet. CDR has been employed in human mobility research with individual user location records. Since the early 2000s, the use of cellular network data for the development of large-scale mobility sensing has been investigated [16].

While CDR has been used to study a variety of transportation issues, such as large-scale urban sensing [17, 18], traffic parameter estimation [19–21], O-D flow estimation [22–24], and land use inference [25, 26], efforts to use CDR data to infer migration trips and model migration flows have been overlooked. By employing CDR data to expand our understanding and modeling of migration flow, this work improves on past studies on the subject. As a result, this research makes two separate contributions: (i) the creation of a heuristic-based approach for inferring migration flows using CDR data; and (ii) the development of trip distribution models for distributing migration flows on a national scale. Our goal is to obtain actionable information from migration flows across the country. These insights can be used by transportation planners to develop better travel demand planning strategies. For example, inferred migration flows can be employed as long-distance journeys in regional and provincial models, and the results of our investigation can be used as an indicator of demographic changes and general migration patterns.

4.2   Methodology

We began our research by extracting a collection of subjects from a mobile phone network dataset (i.e., CDR). A residential location was determined for each subject based on the most commonly used cell tower locations at night. The shift in household locations was then used to infer migration (homes). The migration trip distribution was described using a series of statistical models. Figure 4.1 depicts our analysis process.

4.2.1   Dataset

For this study, we used anonymised CDR data obtained from 1,891,928 mobile phone users in Portugal (about 18% of the population) over a 14-month period (April 2, 2006–June 30, 2007, with half of September and all of October 2006 records missing). The caller ID, callee ID, caller's linked cell tower ID, callee's connected cell tower ID, call duration, and timestamp are all included in each record. The nearest cell tower position is recorded each time a mobile phone user makes or receives a call, i.e., connects to a cell tower.

Individual phone numbers were anonymised by the operator before leaving their storage facilities to protect personal privacy, and were identifiable with a security ID (hash code). There is no information in the dataset about text messages (SMS) or data usage (Internet). There are 6358 cell towers in all, with each serving an average of 14 km^2 in rural areas and 0.13 km^2 in urban areas like Lisbon and Porto. This study only considered cell towers in continental Portugal, not the autonomous territories of Portugal, such as the Azores and Madeira, which are both islands.

Over the course of 14 months, the data includes over 500 million records (cellular network connections). Figure 4.2 illustrates the average amount of recordings

Fig. 4.1 Analysis process

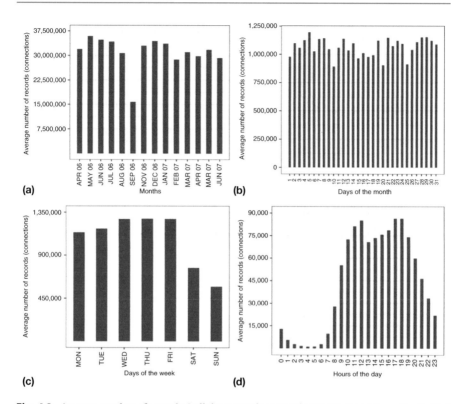

Fig. 4.2 Average number of records (cellular network connections); (**a**) over 14 months (half of September data are missing), (**b**) days of the month, (**c**) days of the week, (**d**) hours of the day

monthly, daily, weekly, and hourly to get a feel of how connection is distributed across time. The missing records are to blame for September's low connectivity. Overall, the networking is simple and straightforward. Summer (May–August) and holiday seasons (November–January) have a high connectivity. From Monday through Friday, connectivity is higher, and during the weekend, it is lower. On an hourly basis, connectivity peaks around midday and again between 5 and 6 p.m. in the evening.

4.2.2 Subjects

To guarantee that we have fine-grained mobility data for our analysis, we chose 538,394 mobile phone users who had at least five connections in each of the 14 months. We needed to identify the site of residence for each subject and ascertain if there was a change because we were interested in migration trips, which is a change of residence.

A residence was determined for each of the 14 months using the same method as Phithakkitnukoon et al. [23], namely identifying the most commonly utilized cell

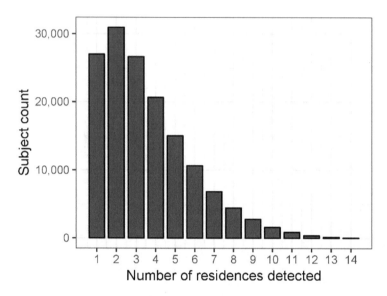

Fig. 4.3 Histogram of the number of residences detected

tower position during the night hours (10 p.m.–7 a.m.) as the approximate location of residence. As a result, only subjects who had connectivity throughout the night hours in each month were evaluated further in our research (i.e., 148,215 subjects). Over the course of 14 months, each of these subjects had a different number of residence locations recognized, ranging from 1 to 14. The detection of only one (same) residence throughout the course of 14 months indicates that there was no change of residence; nevertheless, the detection of multiple residence locations can indicate that there were changes of residence. A histogram of the number of residences found is shown in Fig. 4.3.

There were 27,004 people whose residence remained the same for the whole 14-month period. These subjects can be used to calculate population density in 18 districts in Portugal, as shown in [23]. As a result, Table 4.1 and Fig. 4.4 compare the CDR-based and census population density distributions, as well as a map of Portugal showing the locations of 18 districts. The result is statistically comparable to census data (according to [23]) with a correlation coefficient (R-value) of 0.9401 (as shown in Fig. 4.5). As a result, the observed change of residency can be reasonably used in our further migration analysis.

4.2.3 Migration Flow Inference

We defined migration as a scenario in which a subject's residential location changes from one location to another, with the former location being at least 2 months before the shift and the new location being at least 2 months after the change. It was

Table 4.1 Comparison of population density distributions between CDR-based and census data

District	Census population	CDR-based population
Lisbon	2,250,533	5185
Porto	1,817,117	6274
Setubal	851,258	2640
Braga	848,185	2377
Aveiro	714,200	1301
Leiria	470,930	567
Santarem	453,638	1451
Faro	451,006	393
Coimbra	430,104	781
Viseu	377,653	1005
Viana do Castelo	244,836	562
Vila Real	206,661	1122
Castelo Branco	196,264	391
Evora	166,706	727
Guarda	160,939	170

possible that some subjects may migrate multiple times. However, this study only considered people who had migrated once. All conceivable migration patterns are shown in Fig. 4.6, where A and B represent the former and new residences, respectively. As a consequence, based on our definition of migration, there were 2107 people who migrated once.

There were 1681 intra-district migrations and 426 inter-district migrations among the 2107 migrations discovered. Figure 4.7 depicts the amount of intra-district migrations as well as the flow of inter-district migrations across the country. Figure 4.8 depicts the monthly distribution of these intra-district and inter-district migrations across time, revealing that people relocated within districts primarily during the summer months, particularly May and June. Similarly, inter-district migrations have a high flow in the summer (May to August), but intuitively a low flow in the winter.

Distribution of inter-district migration flow distance in terms of space is depicted in Fig. 4.9. The distance was calculated using the Google Distance Matrix API as a travel distance on the road network, with the closest distance being 35.29 km between Castelo Branco and Guarda. The longest distance between Braganca and Faro is 523.22 km. The average distance traveled is 233.59 km. Figure 4.9 shows two groupings of migration distances centered on the average distance (indicated by a dash line). Migrations to adjacent districts go into the below average-distance category, whereas long-distance migrations, primarily between major cities such as Lisbon and Porto, fall into the above average-distance category.

In the form of an Origin-Destination (O-D) matrix that describes the people's movement from the 'origin,' which is the previous residence, to the 'destination,' which is the new residence, Fig. 4.10 shows a checkerboard plot ranked by origins, i.e., the amount of originated inter-district migrations. For transportation system

Fig. 4.4 Map of Portugal

planning, an O-D matrix is commonly employed. It provides information on traffic volumes, which represent transportation demand. It is a transportation need for migration in this scenario.

The largest outflow comes from Lisbon, followed by Porto, Setubal, and Braga. Setubal, Santarem, and Porto are the top destinations for Lisbon's outflows. Setubal

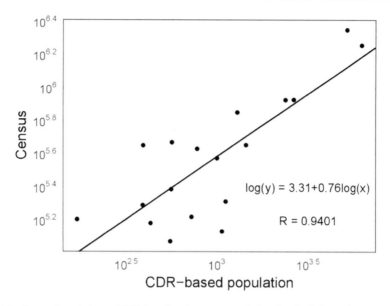

Fig. 4.5 Comparison between CDR-based and census population density information

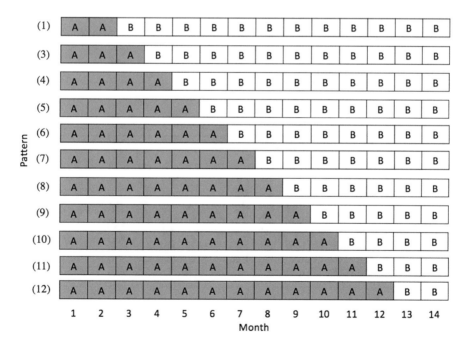

Fig. 4.6 Considered patterns of migration in our analysis, where A and B are the previous and new residence

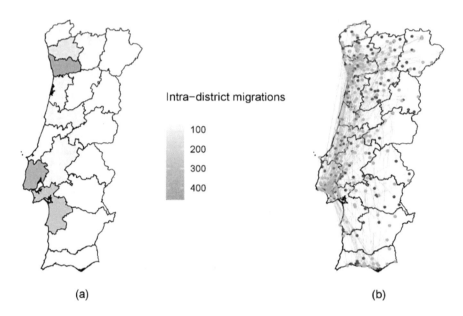

Fig. 4.7 Detected (**a**) intra-district migrations and (**b**) inter-district migrations, where green and red dots indicate the previous and new residential locations

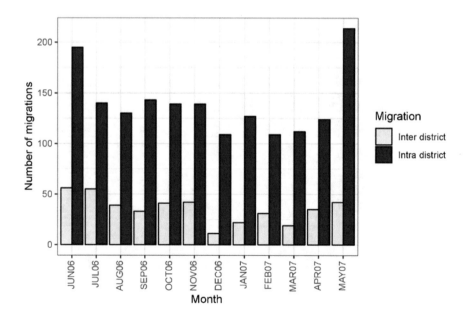

Fig. 4.8 Monthly distribution of intra-district and inter-district migrations

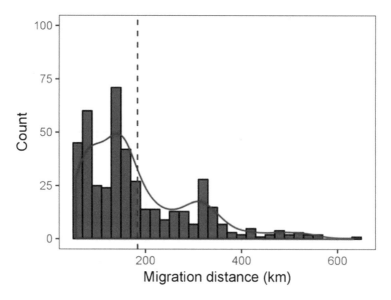

Fig. 4.9 Distribution of the migration distance, where a solid red line and dash line indicate the trend and average

and Santarem are neighbouring areas, while Porto, some 320 km north of Lisbon, is another densely populated district. Lisbon, Braga, and Aveiro are the top destinations for Porto's outflows. Porto is located near the districts of Braga and Aveiro, whereas Lisbon is located to the south of Porto and is the country's capital and largest city.

In terms of migratory inflows, Lisbon is the most popular destination, followed by Porto. Lisbon's biggest inflows are Setubal, Porto, and Santarem, whereas Porto's top inflows are Braga, Lisbon, and Braganca. Braganca is located about 200 km northeast of Porto. The locations of the aforementioned municipalities can be seen in Fig. 4.4.

There were also individuals who migrated within the same district, in addition to the migration movements between districts. There was a total of 1681 intra-district migration flows. The number of intra-district migrant flows, ranked by district, is shown in Fig. 4.11. Surprisingly, Porto has higher traffic than Lisbon, the country's capital and largest city. Setubal is ranked third, while Braga is placed fourth.

4.2.4 Migration Flow Modelling

A trip generation model can be used to estimate the total trip productions and attractions of each district in the case of a travel demand forecasting model. Then, using a trip distribution model, trips (generated by a trip generation model) can be distributed among destinations. In Portugal, there is no comprehensive data on migration flows that may be utilized to create a trip generation model. We

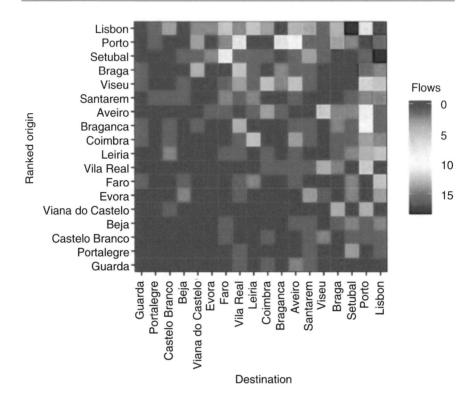

Fig. 4.10 O-D flows of inter-district migrations ranked by the origins (previous residence)

established a simple approach in this work to enlarge the sample migration O-D flows to explain the complete population's migration flow behavior. The total permanent migration trips originating in each district, as well as the total permanent migration trips destined for each district, are then calculated to indicate each district's migration-specific trip generation and attraction functions. We also examined trip distribution models to see how migrant flows in Portugal are distributed.

4.2.4.1 Expansion of Migration Trips

The migration flows derived from the CDR data in the last section are only partial. The next stage is to enlarge the sample migration O-D matrix to represent the complete population's migration flow behavior. An expansion factor based on sampling ratio can be used to convert a sample migration O-D flow to a population migration O-D flow. A simple procedure is developed to calculate: (i) the total permanent migration trips originating from each district (MO_i); and (ii) the total permanent migration trips destined to each district (MD_j).

To compute MO_i, first determine the number of sampled residents in each district. A total of 148,215 users are used, from which we can determine whether they relocated permanently or not. The number of residents sampled is determined by

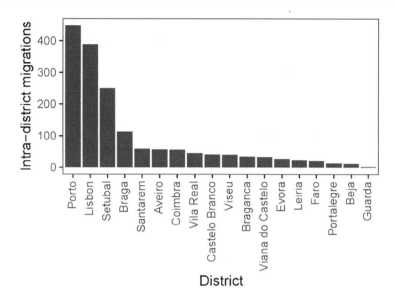

Fig. 4.11 Number of intra-district migrations ranked by district

the number of households discovered in each district. This entails assigning the 148,215 users to each of the districts based on their first-time dwelling location, as determined by CDR data. The second stage is to use the sampled resident data to calculate the total number of migrants from each district (MO_{si}). Then, for each district, an expansion factor (f_i) is calculated as the ratio between the total number of population (people aged 20–59) and the number of sampled users identified as residents of that district based on the CDR data. Finally, MO_{si} and f_i are multiplied to get the overall permanent migration flow (MO_i) from each district. A census taken near the time period of the CDR data is utilized as a secondary source of data describing the population that may be linked to the sample in order to calculate MO_i. We were unable to locate a suitable secondary source of information describing the total migration flow incentives in each district. The model for total permanent migration trips emanating from each district was deemed to be fair. As a result, we maintain control over the total number of permanent migration trips, ensuring that the total number of permanent migration trip sources equals the total number of permanent migration trip destinations.

4.2.4.2 Migration Trip Distribution Modeling

The second phase in the traditional four-step transportation travel demand forecasting model is trip distribution modeling. In order to distribute migration flows in Portugal, we examined three well-known models. The first is the log-linear model, which is based on statistical estimation of flows as a function of many exploratory variables such as origin and destination zone characteristics, as well as the travel cost parameter. The second is the gravity model which generates the most likely set of flows given a set of restrictions that reflect total trip productions and generations. The third model is the radiation model, which is inspired by Stouffer's framework of intervening opportunities [27] and derived from diffusion dynamics. It is completely parameter-free and is based solely on population spatial distribution.

4.2.4.2.1 Gravity Model

The gravity model, which is based on an analogy with Newton's gravitational law, defines the volume of trips made between two places that is proportional to their population but inversely proportional to the travel cost between them. Mathematically, the relationship is defined by Eq. (4.1), as follows.

$$T_{ij} = A_i O_i B_j D_j f(C_{ij}), T_{ij} = A_i O_i B_j D_j f(C_{ij}) T_{ij} = A_i O_i B_j D_j f(C_{ij}), \qquad (4.1)$$

where T_{ij} is the number of undirected flows between district i and district j, O_i and D_j are the total trip ends of districts i and j respectively, and $f(c_{ij})$ is a generalized travel cost function between districts i and j. The two sets of balancing factors $A_i =$

$1/\sum_j B_j D_j f(C_{ij})$ and $B_j = 1/\sum_i A_i O_i f(C_{ij})$, which ensure that the estimates of T_{ij},

when summed across both rows and columns of the matrix equal the known O_i and D_j totals [28]. The popular versions for cost function, $f(C_{ij})$, are negative exponential function ($e^{-\beta C_{ij}}$), inverse power function (C_{ij}^{-n}), and combined function ($C_{ij}^{-n} e^{-\beta C_{ij}}$), where β and n are exponential and power parameters for the cost function, respectively.

The gravity model has been used in economics (e.g., trade) and transportation research for a long time (e.g., mobility). Fisk and Linnemann [29] and Bouchard and Pyers [30] studied international commerce flows and urban trip distribution patterns, respectively, as early implementations of the approach. Later research (e.g., [31, 32]) has shown that the link is still valid. However, the gravity model has some drawbacks, including being deterministic (i.e., it can't account for changes in the number of travelers), systematic predictive inconsistency and discrepancies [33], parameters that aren't always easy to calibrate [34], and a lack of emphasis on potentially important social aspects [35].

4.2.4.2.2 Log-Linear Model

The analysis of values included within contingency tables was done using a log-linear model. Our O-D flow analysis between 18 districts can be thought of as a two-way contingency table. The doubly constrained, multiplicative model in Eq. (4.2) can be equal to the statistical (additive) log-linear model in Eq. (4.3), according to a study by [7].The multiplicative version of the log-linear model describing the entire system can be stated as illustrated in Eq. (4.2) using a notation similar to [6].

$$T_{ij} = \tau \tau_i^O \tau_j^D \tau_{ij}^{OD}, \tag{4.2}$$

where, τ is the overall main effect representing the level of migration, τ_i^O and τ_j^D are the origin and destination 'main effects' represented by categorical variables, respectively. Each of them has 18 levels (the total number of districts in Portugal), τ_{ij}^{OD} is an origin-destination interaction component representing the physical or social distance between districts not explained by the other three components with parameters, where $n = 18$ (i.e., 306 in total for our case, where intra-district migrations are not included). By taking a natural logarithm, the multiplicative log-linear model in Eq. (4.2), can be expressed as a log-linear (additive) model as follows in Eq. (4.3), where $ln(\tau) = \lambda$.

$$ln\left(T_{ij}\right) = \lambda + \lambda_i^O + \lambda_j^D + \lambda_{ij}^{OD} \tag{4.3}$$

4.2.4.2.3 Radiation Model

The radiation model was inspired by Stouffer's paradigm of intervening opportunities [27] and derived from diffusion dynamics. It is completely parameter-free and is based solely on population spatial distribution. The radiation model [33] depicts the number of trips taken between two locations as a job-selection process that includes both job seeking and job selection. Job selection considers the quantity of work options in each location to be proportional to the resident population, whereas job selection evaluates the nearest job with a benefit higher than the best offer available in the resident area.

The model's equation mathematically links the origin population, the destination population, and the total population within the circle centered on the origin region with the distance radius between the origin and destination areas, as described in Eq. (4.4). Finally, the number of trips from the origin to the destination is determined as a part of the origin's total outflows.

$$T_{ij} = T_i \frac{m_i n_j}{\left(m_i + s_{ij}\right)\left(m_i + n_j + s_{ij}\right)}, \tag{4.4}$$

where T_{ij} denotes the number of trips made from origin i to destination j, T_i is total number of trips departing from location i, m_i is the population of area i, n_j is the population of area j, and s_{ij} is the enclosed population in the circle with radius the distance between areas i and j excluding the populations of i and j.

4.2.4.3 Generalized Cost

4.2.4.3.1 Travel Cost Measurements

The cost of traveling between two traffic analysis zones is known as travel cost (i.e., districts in our case). Measuring the travel cost is crucial for evaluating migration flows since it conveys the trip maker's perception of the disutility of travel. Many factors influence the actual cost of a journey, including travel time, distance, fuel consumption, and individual efforts (e.g., physical work involved, willingness, travel period, etc.), making it difficult to measure the actual travel cost. The generalized cost of travel is a value that is calculated by adding all of the monetary and non-monetary costs of a journey together. The generalized cost can be expressed as a linear function of all the trip's monitory and non-monetary costs, weighted by coefficients that try to represent the relative importance as judged by the trip maker [36]. This study does not attempt to estimate these coefficients, therefore three metrics of travel cost based on distance and monetary values are studied separately.

4.2.4.3.1.1 Displacement

Displacement is a linear distance between two geo locations, computed using Haversine formula, given by Eq. (4.5).

$$d = 2R \cdot arcsin\left(\sqrt{cos\left(lat1\right) \cdot coscos\left(lat2\right) \cdot \left(\frac{\Delta lon}{2}\right)}\right), \qquad (4.5)$$

where $\Delta lat = lat2 - lat1$, $\Delta lon = lon2 - lon1$, and R is the Earth's radius (for which we used the average radius of 6371 km).

4.2.4.3.1.2 Road Network Distance

The shortest path between two points in a road network is determined using the Google Maps Distance Matrix API.

4.2.4.3.1.3 Monetary Cost

The monetary cost of commuting from one place in a road network to another is evaluated using a taxi fare, which may be retrieved using the Taxi Fare Finder API.

4.2.4.3.2 Reference Points

To ascertain the origin and destination geolocations from which a relative numerical value may be determined, measuring a journey cost requires points of reference from both zones. The following are three techniques to determine the reference locations that we considered.

4.2.4.3.2.1 District Centroids

The centroid of the district is used as the reference geolocation point for district centroids [37]. Figure 4.12 is an example of using district centroids as the reference points for two districts, denoted by an x ('x').

4.2.4.3.2.2 Farthest Cell Towers

The reference points for the Farthest Cell Tower Sites are the locations of the cell towers that are the furthest away between the two districts. One is in one district, whereas the other is in a different district, as seen in Fig. 4.12, which is marked with a square ('☐').

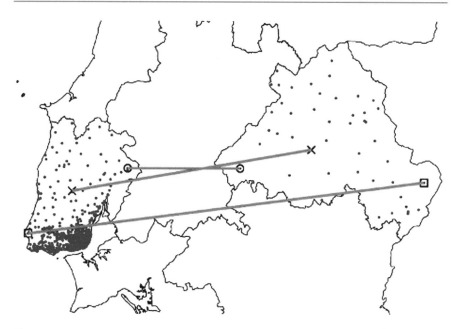

Fig. 4.12 An example of different approaches in determining the reference points for travel cost measure

4.2.4.3.2.3 Nearest Cell Towers

Nearest Cell Tower Locations uses the locations of the cell towers that are closest to each other as reference points, with each tower belonging to one of the two districts, as illustrated in Fig. 4.12 with a circle ('O').

4.3   Results

A trip distribution model is developed to forecast or estimate the number of trips that will be made between two zones, in this instance we have migration flows. There are a variety of models, each with its own description of the incorporated components that are assumed to influence the trip distribution. The standard gravity model, a log-linear model, and a radiation model are all considered here. A total of 27 trip distribution models are estimated using the three modeling methodologies.

4.3.1   Log-Linear model

The log-linear model is used to estimate the first group of models, Models 1.1.1 through 1.3.3. A total of 306 inter-district permanent migration O-D flows were created using an 18-district zoning system. We started with Model 1.1.1's estimation. Categorical variables are associated with an origin component representing the relative 'pushes' from each district (18 categorical variables), and a destination component representing the relative 'pulls' to each district when fitting a doubly restricted log-linear model in Eq. (4.3) (18 categorical variables). An origin-destination interaction is captured using a continuous travel cost variable.

To estimate the log-linear model in Eq. (4.3), we used the Generalized Linear Model (GLM) implemented in R-programming (R Core Team 2016). The observed migratory O-D flows are counts, with a Poisson distribution assumed. Thus, under the GLM framework, the Poisson model assumed that the predicted permanent migration O-D trips follow a Poisson distribution with a mean that is logarithmically connected to a linear combination of the origin and destination specific category variables and the travel cost variable. For illustration, the Model 1.1.1 estimates result is shown in Eq. (4.6), where $\widehat{M1.1.1}_{ij}$ is the Model 1.1.1 estimated permanent migration O-D flow between districts i and j. For the origin-specific and destination-specific categorical variables, the $orig_1$ and $dest_1$ variables were utilized as reference categories, respectively. As a result, the model outputs for the remaining categorical variables should be evaluated in light of the reference categories linked with the Lisbon district. All of the categorical variables' coefficients are negative and significant at the 95% confidence level, indicating that migratory inflow and outflow are low for districts other than the reference group (Lisbon). At the 95% confidence level, the coefficient of the trip cost (lnC_{ij})variable is statistically significant, and its sign is negative.

$$\widehat{M1.1.1}_{ij} = \exp\left(12.346 - 0.997 orig_2 - 1.701 orig_3 - 3.595 orig_4 + \ldots \right. \tag{4.6}$$

$$- 2.555 orig_{18} - 1.786 dest_2 - 1.652 dest_3 - 1.051 dest_4 + \ldots$$

$$\left. - 2.388 dest_{18} - 1.055 \ln C_{ij} \right)$$

For the nine log-linear models, Fig. 2.13 provides a comparison of observed (CDR-based) and estimated inter-district migration visits. The result of the log-linear model provided in Eq. (4.6) is shown in Fig. 4.13a, which is compared to CDR-based migration flows. To examine the linear relationship between the observed visits and the model outputs, we generated a more traditional correlation value. Table 4.2 contains a comprehensive list of R values.

Table 4.2 presents the results of nine log-linear models' estimations. Table 4.2 show that all nine estimated models have relatively high R-values, ranging from 0.8625 to 0.8835, indicating a strong fit between observed and estimated migratory O-D fluxes. Table 4.2 shows the travel cost (lnC_{ij}) parameters of the nine doubly constrained log-linear models. The trip cost parameter's coefficient is negative in all nine models, which makes sense because the travel cost variable indicates the expense of migration between the origin and destination districts, and the attractiveness of a destination district diminishes as the journey cost rises. Between the models that were estimated using a similar trip cost specification, there is a discrepancy in the value of the travel cost parameter. In Models 1.1.1 and 1.2.1, for example, the district centroid is utilized as a reference point to calculate the travel cost (distance)

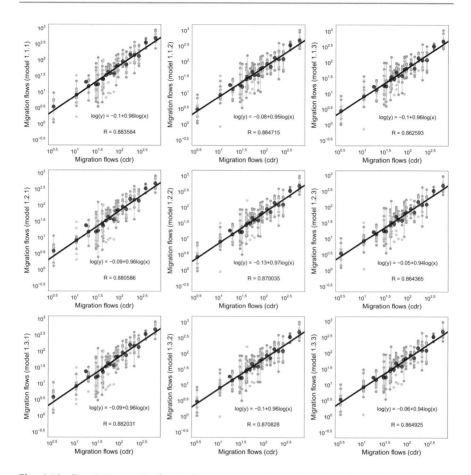

Fig. 4.13 Correlation results for log-linear models using various techniques. For each pair of districts, grey points represent scatter plots. If the fitted regression line falls between the 9th and 91st percentiles in that bin, the box is green; otherwise, it is red. The mean number of estimated migratory flows in the bin is represented by the black circles with cross lines

between the origin and destination districts. Model 1.1.1's trip cost parameter (-1.055) is, however, higher than Model 1.2.1's travel cost parameter (-1.227). One of the key reasons is that Model 1.2.1 uses the real journey distance on a road network, resulting in a higher average travel cost than Model 1.2.1. As a result, Model 1.2.1's higher average travel cost makes it more susceptible to the distance-decay effect, and the migratory interaction between two districts decreases as the trip cost (measured in travel distance) between them increases.

Table 4.2 Result summary of the log-linear models

Model	Travel cost	Reference point	Travel cost parameter (lnC_{ij})	R-value	p-value	F-test[a]
1.1.1	Displacement	Centroid	−1.055	0.883584	$<10^{-16}$	1082
1.1.2	Displacement	Farthest	−0.856	0.864715	$<10^{-16}$	901.1
1.1.3	Displacement	Nearest	−0.694	0.862593	$<10^{-16}$	883.8
1.2.1	Road distance	Centroid	−1.227	0.880586	$<10^{-16}$	1050
1.2.2	Road distance	Farthest	−0.900	0.870035	$<10^{-16}$	946.8
1.2.3	Road distance	Nearest	−0.811	0.864365	$<10^{-16}$	898.2
1.3.1	Taxi fare	Centroid	−1.285	0.882031	$<10^{-16}$	1065
1.3.2	Taxi fare	Farthest	−1.042	0.870828	$<10^{-16}$	954
1.3.3	Taxi fare	Nearest	−0.927	0.864925	$<10^{-16}$	902.8

[a]F-statistic = 3.91

Table 4.3 Result summary of the gravity models

Model	Travel cost	Reference point	R-value	p-value	F-test[a]
2.1.1	Displacement	Centroid	0.883583	$<10^{-16}$	1082
2.1.2	Displacement	Farthest	0.864716	$<10^{-16}$	901.1
2.1.3	Displacement	Nearest	0.862593	$<10^{-16}$	883.8
2.2.1	Road distance	Centroid	0.880587	$<10^{-16}$	1050
2.2.2	Road distance	Farthest	0.870036	$<10^{-16}$	946.8
2.2.3	Road distance	Nearest	0.864365	$<10^{-16}$	898.2
2.3.1	Taxi fare	Centroid	0.882031	$<10^{-16}$	1065
2.3.2	Taxi fare	Farthest	0.870829	$<10^{-16}$	954
2.3.3	Taxi fare	Nearest	0.864925	$<10^{-16}$	902.8

[a]F-statistic = 3.91

4.3.2 Gravity Model

The gravity model, despite its flaws, does not impose a substantial computational overhead and is easily scalable with real-world populations utilizing only a few data variables. The results of nine doubly restricted gravity models for permanent migration flows are shown in Table 4.3. In order to reasonably duplicate the observed (benchmark) migration flow pattern, fitting the gravity model in Eq. (4.1) necessitates determination of its parameters. A_i, B_j, and a travel cost parameter are among the parameters on the list. In our case, the parameters A_i and B_j are calibrated during the gravity model calculation, whereas the travel cost parameter is taken from

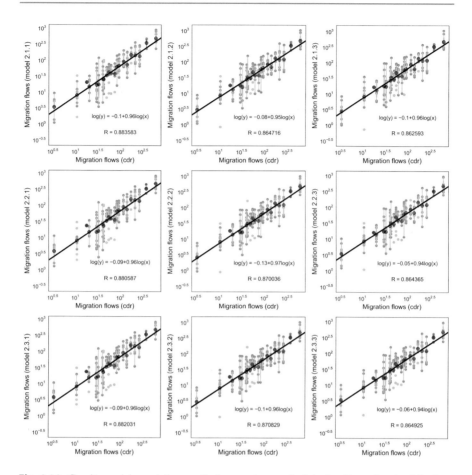

Fig. 4.14 Gravity model correlation results from various methodologies. For each pair of districts, grey points represent scatter plots. If the fitted regression line falls between the 9th and 91st percentiles in that bin, the box is green; otherwise, it is red. The mean number of estimated migratory flows in the bin is represented by the black circles with cross lines

Table 4.2 (the corresponding log-linear model estimates). Model 2.1.1, for example, is estimated using 18 A_i parameters, 18 B_j parameters, and 1 trip cost parameter, such as -1.055 from Eq. (4.6).

A regression line was fitted to each model, and the correlation (R-value) between the actual or observed migration flows (i.e., CDR based) and the gravity model's predicted or expected migration flows was measured. Figure 4.14 depicts the outcome, which is described in Table 4.3.

The Model 2.1.1, a gravity with displacement distance as the travel cost and district centroids as the reference points, has the highest correlation (R = 0.8835, $p < 10^{-16}$), followed by the Model 2.3.1, a gravity with taxi fare as the travel cost and district centroids as the reference points (R = 0.8820, $p < 10^{-16}$). Model 2.1.3, gravity with displacement and nearest cell tower locations as reference points (R = 0.8625, $p < 10^{-16}$) and Model 2.2.3, gravity with road distance as its travel cost and nearest cell tower locations as reference points (R = 0.8643, $p < 10^{-16}$) are among the models with the lowest correlations. All nine models have similar high correlation values, ranging from 0.8835 to 0.8625. Models 2.1.1, 2.3.1, 2.2.1, 2.3.2, 2.2.2, 2.3.3, 2.1.2, 2.2.3, and 2.1.3 are ranked in order from highest to lowest based on correlation values. When the average of each travel cost parameter is taken into account, the displacement (0.8820) comes in top, followed by road distance (0.8685), and finally taxi fee (0.8638). The ranking is district centroid (0.873042), farthest cell towers (0.8721), and nearest cell towers (0.8721) when using the reference point approach (0.8693). In conclusion, the best fit for the gravity model is the combination of displacement distance as the travel cost and district centroid. When it comes to individual trip costs and reference points, the district centroid and taxi fare are the best options.

4.3.3 Radiation Model

The log-linear and gravity models were used to determine the distribution of migration-related trips. We also compared the CDR-based migration flow to radiation model predictions, which are based on important migration factors including firm establishment data, employment data, and so on. The radiation model is primarily based on population or mass-based parameters (m_i, n_j, s_{ij}), which represent geographic attractiveness, which is thought to influence trip distribution. By altering the source of data for these factors, we were able to predict the radiation model. On the basis of nine radiation models, various data sources are employed to characterize the attractiveness of the place. Models 2.1 and 2.2 are based on census data, with the former using the district's total population and the latter using the district's population per km^2. Models 2.3 and 2.4 are based on data collected from the Google Places API, which gives lists of local companies based on the total number of places (local businesses) and the number of places per km^2, respectively. Model 2.5 is based on the Global Competitiveness Index (GCI), a 12-pillar competitiveness index that measures economic performance [38]. On September 30, 2006, the Pblico newspaper published GCI by city data of Portuguese district capitals. Model 2.6 is based on employment data from the census [39], which shows the number of people aged 15 and up who worked for a wage or salary during the reference year. Model 2.7 is based on unemployment statistics from the census [39], which shows the number of people aged 15–74 who did not have a job or were working during the reference year. Model 2.8 is based on census data on resident deaths, which reports the number of residents who died or disappeared permanently during the reference year. Finally, Model 2.9 employs data from the census on private households to calculate the number of groups of persons living in the same unit who are linked to each other (by law or 'de facto') during the reference year. The reference year for all census data used here was 2011, as it was the most recent census data available at the time of our CDR data collection.

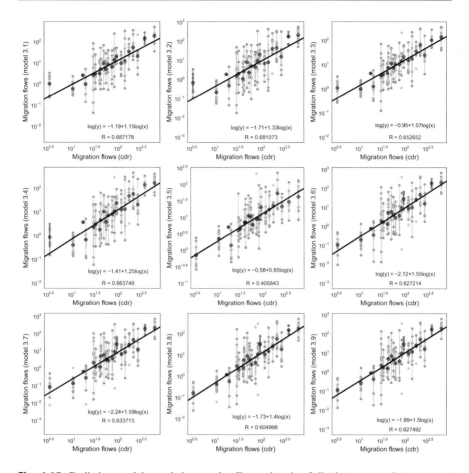

Fig. 4.15 Radiation model correlation results. For each pair of districts, grey points represent scatter plots. If the fitted regression line falls between the 9th and 91st percentiles in that bin, the box is green; otherwise, it is red. The mean number of estimated migratory flows in the bin is represented by the black circles with cross lines

A regression line was fitted to each radiation model, and the correlation (R-value) between the real or observed migration flows (i.e., CDR based) and the model's estimate was measured. Figure 4.15 depicts the resulting regression, and Table 4.4 summarizes the statistical values.

Model 3.1, the district population-based model, has the highest correlation value (R = 0.6871, $p < 10^{-16}$), followed by Model 3.2, the distance population per km^2-based model (R = 0.6810, $p < 10^{-16}$), and the GCI-based model, surprisingly, has the lowest correlation value (R = 0.4058, $p < 10^{-16}$). Models 2.1, 2.2, 2.4, 2.3, 2.7, 2.9, 2.6, 2.8, and 2.5, in order of highest to lowest correlation values, are ranked from highest to lowest. The population and local business models appear to have a better relationship, with a relatively high correlation value (R is above 0.65).

Table 4.4 Result summary of the radiation models

Model	Population or mass terms (m_i, n_j, s_{ij})	R-value	p-value	F-test[a]
3.1	District population	0.687178	$<10^{-16}$	272
3.2	District population per km^2	0.681073	$<10^{-16}$	263
3.3	Number of local businesses	0.652652	$<10^{-16}$	225.6
3.4	Number of local businesses per km^2	0.663748	$<10^{-16}$	239.4
3.5	Global Competitiveness Index	0.405843	$<10^{-13}$	59.94
3.6	Employment	0.627214	$<10^{-16}$	197.2
3.7	Unemployment	0.633713	$<10^{-16}$	204
3.8	Death of residents	0.604998	$<10^{-16}$	175.5
3.9	Private households	0.627492	$<10^{-16}$	197.4

[a]F-statistic $= 3.91$

The findings of various techniques to define the mass terms for the radiation model do not outperform the gravity and log-linear models. In fact, the obtained correlation values, which represent the model's fit to CDR-based migration trips, are significantly lower than those of the other two models. This could be because the radiation model is only focused on population spatial distribution, and thus the journey cost, which is also one of the most important influencing elements in capturing an origin-destination interaction, is not taken into account.

The purpose of this migration trip distribution modeling is to show that our intuitive method to migration trip inference is effective and can be reasonably characterized by well-known models. As can be seen, the resulting estimates differ depending on the model's characterisation of the integrated elements that are supposed to affect migration.

4.4   Conclusion

Our lives have become inextricably linked to our mobile phones. It has also become our own behavioral sensor, thanks to its sensory components and communication technology. Communication records can be utilized to better understand behaviors and provide in-depth information on people and city features when seen as a whole. We were able to reasonably infer about migration trips (or change of residence) for which multiple trip distribution models were investigated to mathematically explain such human mobility by employing the technique of using a large-scale mobile phone network data (CDR: call detail records) in a case study discussed this chapter. We described a methodology, which includes data pre-processing, subject selection, and migration trip inference, from which we discovered some interesting findings, including intra/inter-district migration flow characteristics, migration distance distri- bution, and migration origin-destination (O-D) movements. The log-linear model, traditional gravity model, and recently developed radiation model were investigated for migration trip modeling, with distinct approaches utilized in setting parameters for each model. Displacement, road network distance, and monetary cost were all included as alternative trip costs in the log-linear and gravity models, while the reference points were determined experimentally as district centroids, farthest cell tower locations, and nearest cell tower sites. The population or mass parameters for the radiation model were defined experimentally as district population, number of local enterprises, Global Competitiveness Index, employment, unemployment, resident death, and private households. The discussed findings and methods add to the body of knowledge in the field of human mobility research.

The research discussed in this chapter does, however, have certain limitations. To begin with, only subjects who had previously migrated were inferred and considered in this investigation. With the examination of numerous migration scenarios, the results produced here could potentially vary. Second, due to the intra-zonal travel distance, only inter-district migration was taken into account in this study's trip distribution modeling. In a future study, it would be worthwhile to investigate intra-district migration trip distribution modeling by examining various methods for estimating the trip distance. Finally, there was a lack of ground truth for the inferred migration trips, due to data unavailability. Despite the lack of a ground truth, the migration inference was validated using a reasonable interpretation of the census-based population density evaluation. Furthermore, the estimated migratory flow in this study is based on only 18% of the population, which does not represent the whole population. A population sample of each district was employed in this study to expand the sample migration O-D flow to a population migration O-D flow. To adequately explain the composition of the sampled data, future studies should seek to also include socioeconomic and demographic information of sample subjects.

References

1. Song C, Qu Z, Blumm N, Barabási AL. Limits of predictability in human mobility. Science. 2010;327(5968):1018–21.
2. Mervyn P. Factors influencing migration and population movements – part 1. Future directions international. 2014. [Online]. https://www.futuredirections.org.au/publication/factors-influencing-migration-and-population-movements/.
3. Charlene Rohr FT, Fox J, Daly A, Patruni B, Patil S. Modelling long-distance travel in the UK. 2010.
4. European Commission. Roadmap to a single European transport area – towards a competitive and resource efficient transport system (Commission staff working document accompanying the White Paper). 2011.
5. Mcnally MG. The four step model. Handb Transp Model. 2007.
6. Stopher PR, Greaves SP. Household travel surveys: where are we going? Transp Res Part A Policy Pract. 2007;21(5):367–81.
7. Raymer J. The estimation of international migration flows: a general technique focused on the origin-destination association structure. Environ Plan A Econ Sp. 2007;39(4):985–95.
8. Blumenstock JE. Inferring patterns of internal migration from mobile phone call records: evidence from Rwanda. Inf Technol Dev. 2012.
9. Zagheni E, Garimella VRK, Weber I, State B. Inferring international and internal migration patterns from Twitter data. In: Proceedings of the 23rd international conference on world wide web; 2014. p. 439–44.

10. Massey DS, Capoferro C. Measuring undocumented migration. Int Migr Rev. 2004;38(3): 1075–102.
11. Beegle K, de Weerdt J, Dercon S. Migration and economic mobility in Tanzania: evidence from a tracking survey. Rev Econ Stat. 2011.
12. Demissie MG. Combining datasets from multiple sources for urban and transportation planning: emphasis on cellular network data. Coimbra: Coimbra Univ.; 2014.
13. Shen L, Stopher PR. Review of GPS travel survey and GPS data-processing methods. Transp Rev. 2014.
14. Cuttone A, Lehmann S, González MC. Understanding predictability and exploration in human mobility. EPJ Data Sci. 2018.
15. Phithakkitnukoon S, Horanont T, Witayangkurn A, Siri R, Sekimoto Y, Shibasaki R. Understanding tourist behavior using large-scale mobile sensing approach: a case study of mobile phone users in Japan. Pervasive Mob Comput. 2015;18.
16. Caceres N, Wideberg JP, Benitez FG. Review of traffic data estimations extracted from cellular networks. IET Intell Transp Syst. 2008;2(3):179–92.
17. Calabrese F, Colonna M, Lovisolo P, Parata D, Ratti C. Real-time urban monitoring using cell phones: a case study in Rome. IEEE Trans Intell Transp Syst. 2011.
18. Ratti C, Sevtsuk A, Huang S, Pailer R. Mobile landscapes: graz in real time. Locat Based Serv TeleCartography. 2007;(October 2005):433–44.
19. Bar-Gera H. Evaluation of a cellular phone-based system for measurements of traffic speeds and travel times: a case study from Israel. Transp Res Part C Emerg Technol. 2007.
20. Demissie MG, de Almeida Correia GH, Bento C. Intelligent road traffic status detection system through cellular networks handover information: an exploratory study. Transp Res Part C Emerg Technol. 2013.
21. Liu HX, Danczyk A, Brewer R, Starr R. Evaluation of cell phone traffic data in Minnesota. Transp Res Rec J Transp Res Board. 2008;2086(1):1–7.
22. Demissie MG, Phithakkitnukoon S, Kattan L. Trip distribution modeling using mobile phone data: emphasis on intra-zonal trips. IEEE Trans Intell Transp Syst. 2019;20(7):2605–17.
23. Phithakkitnukoon S, Smoreda Z, Olivier P. Socio-geography of human mobility: a study using longitudinal mobile phone data. PLoS One. 2012;7(6):e39253.
24. Calabrese F, Di Lorenzo G, Liu L, Ratti C. Estimating origin-destination flows using mobile phone location data. IEEE Pervasive Comput. 2011.
25. Toole JL, Ulm M, González MC, Bauer D. Inferring land use from mobile phone activity. In: Proceedings of the ACM SIGKDD international conference on knowledge discovery and data mining; 2012. p. 1–8.
26. Demissie MG, Correia G, Bento C. Analysis of the pattern and intensity of urban activities through aggregate cellphone usage. Transp A Transp Sci. 2015.
27. Stouffer SA. Intervening opportunities: a theory relating mobility and distance. Am Sociol Rev. 1940;5(6):845–67.
28. Wilson AG. The use of entropy maximising models, in the theory of trip distribution, mode split and route split. J Transp Econ Policy. 1969.
29. Ball RJ, Linnemann H. An econometric study of international trade flows. Econ J. 1967;77 (306):366–8.
30. Bouchard RJ, Pyers CE. Use of gravity model for describing urban travel: an analysis and critique. Highw Res Rec. 1965;88(88):1–43.
31. Abdel-Aal MMM. Calibrating a trip distribution gravity model stratified by the trip purposes for the city of Alexandria. Alex Eng J. 2014;53(3):677–89.
32. Lenormand M, Bassolas A, Ramasco JJ. Systematic comparison of trip distribution laws and models. J Transp Geogr. 2016;51:158–69.
33. Simini F, González MC, Maritan A, Barabási AL. A universal model for mobility and migration patterns. Nature. 2012;484:96–100.
34. Gargiulo F, Lenormand M, Huet S, Espinosa OB. Commuting network models: getting the essentials. J Artif Soc Soc Simul. 2012;15(2). https://doi.org/10.18564/jasss.1964.

35. Kraft S, Blazek J. Spatial interactions and regionalisation of the Vysocina region using the gravity models. Acta Univ Palacki Olomuc – Geogr. 2012;43(2):65–82.
36. de Dios Ortúzar J, Willumsen LG. Modelling transport. Chichester, UK: Wiley; 2011.
37. Demissie MG, Phithakkitnukoon S, Kattan L. Understanding human mobility patterns in a developing country using mobile phone data. Data Sci J. 2019;18(1):1–13.
38. Sala-i-Martin X, Artvadi E. The global competitiveness index. Global economic forum. 2004.
39. Francisco Manuel dos Santos Foundation. Database of contemporary Portugal. PORDATA. https://www.pordata.pt/en/Municipalities/Search/5/.

Inferring Social Influence in Transport Mode Choice Using Mobile Phone CDR Data

5

Abstract

Previous chapters have shown how mobile network data can be used to infer travel generation and trip distribution. We continue to explore further in this chapter into how to utilize the mobile network data in making inferences about transport mode choice and its social influence. This chapter focuses on social influence in terms of ego-network effect in commuting mode choice, for which a longitudinal mobile phone data that includes both location and communication records is investigated. Methods for inferring social tie strength and transport mode as well as a framework for analyzing social influence in transport mode choice are discussed. The findings reveal that a person's strong relationships are more essential in determining whether or not driving is the person's preferred mode of transportation, whereas weak ties are more relevant in determining whether or not public transportation is the person's preferred mode of transportation. The data also shows that social ties that are geographically nearby have a greater influence on commuting mode choice than those that are farther away. In the case of public transportation, accessibility distance is also a deciding factor. As the access distance increases, the percentage of people who use public transportation decreases. Furthermore, the social network has been found to influence commute mode choice, with the likelihood of choosing a given mode increasing as the percentage of social ties who choose that mode increases. The content discussed in this chapter reflects the idea, motivation, and thinking process in our original work done by Phithakkitnukoon et al. (EPJ Data Sci. 2017;6(11); Soc Netw Anal Min. 2016;6(1)).

Keywords

Call detail records · Social influence · Social tie strength · Social network analysis · Transport mode inference · Transport mode choice · Commuting trip

5.1   Motivation and State of the Art

A worldwide phenomena has been the occurrence of a variety of trends such as urbanization, globalization, resource shortages, and technological breakthroughs. These shifts have an impact on how planners approach problems and evaluate solutions. The field of transportation is undergoing a paradigm shift as well, with a focus on using an all-encompassing multimodal approach and demand management solutions to reduce private car dependence and improve the efficiency and sustainability of public transit systems [1].

Until recently, the use of private vehicles in developed countries had been steadily increasing. To meet the increased vehicle traffic caused by private car growth, public authorities took a fiscally unsustainable approach, such as expanding the urban road network. In recent years, issues such as population aging, rising fuel prices, urbanization, increased health and environmental impacts, and changing consumer preferences have resulted in a peak in private car travel [2], which has sparked an increase in demand for alternate modes of transportation [1].

A variety of transportation demand management strategies have been investigated to see how effective they are at reducing private car use. Some of these measures are intended to improve the attractiveness of alternative transportation modes, such as providing free bus travel for a specified time period, and others are intended to limit private car use, such as temporary changes to infrastructural conditions, e.g. road closures [3], however, recent experiences show that people are willing to use alternative modes such as walking, bicycling and public transit. To address the growing demand for alternate modes of transportation, service quality in terms of convenience, comfort, cost, and integration must be improved. This cannot be done without first understanding the needs of the travelers and their preferences for alternative modes of transportation. Mode choice modeling is one area that could be improved. It's important for predicting future growth for each mode, as well as identifying factors that influence how people use each mode and how they switch from one to another.

If mode choice models are based on aggregated zonal (and inter-zonal) data, they can be aggregated. It can also have disaggregated models if they are based on household and/or individual data. Demand, according to disaggregate models, is the product of multiple decisions made by each individual passenger. Multinomial logit and nested logit, the two most prominent discrete-choice models, have been used to explore factors influencing travel behavior [4, 5]. The discrete-choice model's theoretical foundation is that people's travel decisions are based on the utility maximization principle, or the relative attractiveness of competing alternatives. An individual's probability as a function of any number of factors that describe the alternatives is predicted by a discrete-choice model.

Discrete-choice models can be based on either observed behavior or hypothetical choice surveys that contain datasets of three main categories, which are believed to be important to influence mode choice: (i) characteristics of the trip maker—car ownership, driver's license, household structure, income, and so on; (ii) characteristics of the journey—trip purpose, time of day, and whether the trip is taken alone or with others; and (iii) characteristics of the transportation facility—monetary cost components, in-vehicle and out-of-vehicle times, parking, comfort and convenience, safety, and so on [5].

Discrete-choice models are based on economic theories and assume that consumers make rational decisions when choosing between forms of transportation. The essential premise is that people try to maximize the utility obtained from the varied attributes of the options and make reasonable decisions. However, some social aspects that influence people's transportation mode choice are missing from these economic theories and approaches. The question of social influence in transportation is receiving more attention [6]. Social network theory recognizes the relevance of social contagion and spreading processes in the decision-making process, and social relationships can have a direct impact on various decisions, such as mode of transportation. For example, when more members of an ego-network use a specific means of transportation, that member is more likely to use that mode as well, which could be due to conformity, social pressure, or an imitation mechanism. As a result, threshold models of influence [7] depict others' impact as a function of the proportion of other people who adopt a new behavior. Network homophily, on the other hand, may be responsible for the imitation process, in which people adjust their activities to match them more closely with the behaviors of their friends [8, 9].

In this chapter, we use mobile phone network data for two purposes: (i) inferring transportation mode, and (ii) investigating the impact of social network variables on transportation mode choice. Our analysis focuses on commuting trips, i.e., trips between a person's home and their place of employment. A study done by Kowald et al. [10] on the impact of social networks on leisure travels is a complementary work and recommended for further reading.

There have been studies in recent years that looked into the use of mobile network data for various purposes. These research explored a variety of topics, including traffic estimation [11–14] and origin-destination flow estimation [15–19]; travel demand analysis [20–22]; land-use detection [23–25]; the interaction between users' mobility, location, and the apps they use [26]; and place-related context inference [27, 28]. There is also a substantial body of literature that examines the use of cellular network data in a variety of contexts, including incident and traffic management [29], urban sensing [30]; social networks, security, and privacy concerns [31]. The primary goal of this research is to investigate the ability of mobile phone data to infer transportation mode and, in particular, to examine the impact of social network variables on transport mode choices. The remainder of this literature review section will be devoted to a review of past studies that have used mobile phone data to analyze social networks.

Political polls [32], panic stampedes [33], financial markets [34], cultural markets [35], aid campaigns [36], product rating [37], and answering inquiries [38] are all examples of social influence in common collective decision processes. Some of these collective decisions can trap a population in a suboptimal condition, such as a financial bubble caused by the herding behavior of financial operators [39]. Alternatively, they may influence a system in a beneficial direction, such as increasing tax compliance rates [40] or accelerating weight reduction [41]. However, quantifying how people perceive and respond to social influence is necessary for understanding how such collective decisions are produced, evaluating their benefit to the public, and even directing their results.

5.1.1   Social Influence on Travel Behavior

In transportation studies, the idea that social activities can influence travel decisions is not new. Salomon [42], for example, recognized the desire to belong as a motivator for travel. Time geography research (e.g., [43–46]) has long stressed the importance of social contact, in the form of coupling restrictions, as a fundamental determinant of travel. There is evidence that there is a link between social engagement and the amount of time spent traveling. Harvey and Taylor [47] found that people who work from home spend less time with others in a research based on time-use Canadian data from 1999. Their findings also revealed that these employees spend only 17% of their time (while awake) with others, compared to 50% for those in a traditional job. People who had minimal social engagement with others tended to travel more, according to Harvey and Taylor. These findings imply that working from home does not necessarily minimize travel for everyone, but rather changes the purpose of travel, which is consistent with Arentze and Timmermans' study [48]. Harvey and Taylor [47] addressed the importance of gaining a better knowledge of social contact, and in particular, recognizing the social connection, based on their findings. Previous transportation research has looked into the impact of social organization (including household), socio-demographics, and gender on travel behavior, often at the aggregate level (e.g., [49–52]) and at the individual level (e.g., [53, 54]), but only taking into account the individual's characteristics and not his or her connections to others. Despite this, there has been relatively little effort done to operationalize the impact of these connections, and more especially, the decisions made by others, on individual decision making. Ben-Akiva and Lerman [55] (page 33) noted that "(by) considering a group of persons as a single decision maker... it is possible to abstract partially the complex interactions within... a household or firm," Although this strategy, which is widely used in travel behavior research, provides a good first-hand approximation to complex problems, it overlooks essential components arising from interpersonal interactions. Recent research, on the other hand, has begun to address some of the added complexity that comes with dealing with social interdependencies, as evidenced by the small but expanding literature on social ties and interactions in travel decision making (e.g., [56–59]). Other factors of social influence, such as social tie strength and socio-geography [60] continue to be disregarded in the analysis of travel behavior, despite a small number of cases in this area of research.

5.1.2 Mobile Sensing Approach in Behavior Analysis

The modern mobile phone is not only a means of communication, but also a new gateway for monitoring human activity—a new sensor-networking paradigm that includes humans as part of its sensing infrastructure. Researchers can gather and analyze human behavior on a large scale using the mobile sensing approach. One of the trendiest areas in contemporary research is human mobility. Despite the prevalent assumption that our behaviors are random and unpredictable, Song et al. [61] investigated the mobility patterns of anonymous mobile phone users and determined that human movement follows surprisingly regular patterns and is 93% predictable. Their findings are consistent with those of González et al. [62], who found that while most people travel only short distances and only a few travel hundreds of miles on a regular basis, they all follow a similar pattern regardless of time or distance, and they have a strong desire to return to places they've visited previously. Human mobility's statistical qualities have been researched [63] and applied to applications such as location prediction [64] and interurban analysis [65]. Understanding human movements, for example, is beneficial to epidemiology. Wesolowski et al. [66] investigated how human travel patterns contribute to Malaria spread. Wang et al. [67] demonstrated that human mobility can be used to model fundamental spreading patterns that characterize a mobile virus outbreak. The mobile sensing method has also been employed in social network structure research (e.g., [68, 69]). The scaling ratio in social structure was identified by Phithakkitnukoon and Dantu [70]. Their study reveals that social structure can alter over time due to a variety of factors such as migrations [71] or behavior adaptation, as reported in a study by Eagle et al. [72] in which individuals adjust their communication patterns to improve similarity with their new social context. Later research by Eagle et al. [73] found a link between social variety and economic development. As a result, social connection is vital not just for increasing one's well-being but also for improving one's financial situation. Weather can influence social contacts, according to Phithakkitnukoon et al. [74], and geographical distance is a constraint for social interactions, according to Onnela et al. [75] and Krings et al. [76]. Interacting with social relationships might sometimes appear to be a sign of face-to-face meeting travel [49, 77]. De Domenico et al. [78] showed that information acquired from social links can aid improve individual location prediction. Phithakkitnukoon et al. [59] established links between people's travel scope and their social relationships' locations. As a result, it's critical to comprehend the relationship between social networks and human mobility. However, there has been relatively little research into this relationship. This study aims to investigate and identify the dimensions and characteristics of this type of relationship.

5.2 Methodology

5.2.1 Subject Selection

In this study, we once again make use of an opportunistic sensing approach for real-life behavior monitoring by having human subjects being part of sensing infrastructure. Telecom operators keep records of their mobile phone users' usage logs for billing purposes. Caller ID, callee ID, caller's connected cell tower ID, callee's connected cell tower ID, duration of the call, and timestamp are all included in these usage logs, also known as call detail records (CDRs). The nearest cell tower position is recorded each time a mobile phone user makes or receives a call, i.e., connects to a cellular network. As a result, CDR data gives fine-grained longitudinal information regarding an individual's mobility and sociality collectively. We took use of the fact that this type of data can provide such a deep analysis of mobility and sociality behavior in this study. Over the course of a year, we used anonymized CDR data from mobile phone users in Portugal.

Individual phone numbers were anonymised by the operator before leaving their storage facilities to protect personal privacy, and were identifiable with a security ID (hash code). There is no information in the dataset about text messages (SMS) or data usage (Internet). There are a total of 6509 cell towers, each of which serves an average of 14 km^2 in rural areas but just 0.13 km^2 in major areas like Lisbon and Porto.

We started with a random sample of 100 mobile phone users (whose locations were recorded at least five times each month over a 1-year period) as the ego-subjects and followed the approach of Onnela et al. [68] to collect their alters (i.e., social ties) based on reciprocal conversations. This snowball sampling-style subject recruitment yielded 5305 alters, bringing the total number of individuals to 5405 for the study. This is an egocentric network with the depth of 1.

Each ego-subject spent 307.12 min on the mobile phone each month (roughly 18 min daily) and was linked to the cellular network 307.12 times (calls) monthly (about ten calls daily) across 168.65 different cell tower locations (14.05 cells monthly) on average across 1 year of observation. In Fig. 5.1a–d, histograms of the number of social ties, call duration, call frequency, and mobility (number of different cell sites visited) are shown, respectively.

5.2.2 Residence and Work Location Inference

We were interested in commuting mode choices in this research, which includes travel between one's home and place of work (or study), which accounts for a major fraction of all person travels. As a result, our initial duty was to determine where each ego-subject lived. Since mobile phone users' location information is at the cell tower level, we approximated each ego-subject's place of residence by using our earlier and successful method of identifying residence location as the location of the most commonly used cell tower during the night (10 pm–7 am). This method has

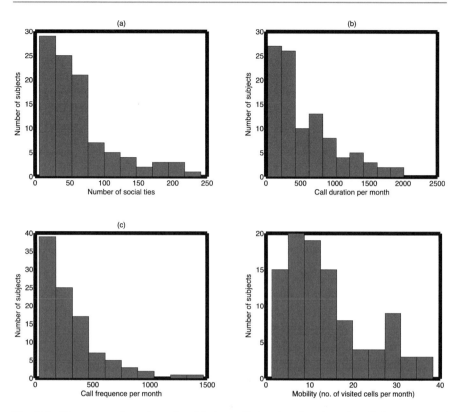

Fig. 5.1 Histograms of number of social ties, call duration (in minutes), call frequency, and mobility

been confirmed against actual census data in a prior work in which we compared the fraction of the estimated population density based on our residence location estimation with the actual population density across all 308 municipalities in Portugal [60]. Figure 5.2 depicts our 100 ego-subjects' predicted dwelling locations overlaid on the Portuguese road network. Our ego-subjects' dwelling locations are scattered geographically over the country, but densely grouped in urban regions such as Lisbon and Porto—intuitively, in line with the country's overall population density distribution.

Similarly, we used this method to estimate each ego-subject's workplace as the location of the cell tower with the highest degree of call activity during typical business hours (9 a.m.–5 p.m.) on weekdays. Figure 5.2b depicts our ego-subjects' estimated housing and workplace locations. The residence is represented by a red cross, while the workplace is represented by a blue circle, with a line connecting the two to symbolize a commuting flow. As commuting flows are denser in cities, Fig. 5.3 shows a zoomed-in version of commuting flows in Lisbon and Porto. (Note that when weekdays and weekends are included, the findings are identical.)

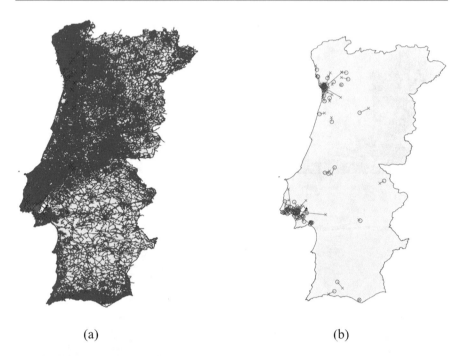

(a)   (b)

Fig. 5.2 (**a**) Estimated ego-subjects' residential locations (indicated with red crosses) overlaid on the road network. (**b**) Estimated ego-subjects' housing and workplace locations (indicated with a red cross and a blue circle, respectively), as well as commute flows (marked with lines)

We calculated the commuting distance for each ego-subject using the estimated residence and work locations. The average distance traveled for commuting was 10.84 km (median distance was 7.12 km). Among the ego-subjects, the longest commuting distance is 48.44 km, while the shortest is 0.49 km. Figure 5.4 shows a histogram of commuting distances.

5.2.3   Social Tie Strength Inference

In addition to location data, CDRs include call records that contain information about an individual's social engagement in the cellular network, which can be used to extract social network information. The caller IDs connected with each ego-subject through reciprocal calls were used to identify the ego-subject's social network. In other words, anyone who had called or received calls to/from the ego-subject was considered the ego-subject's social tie (or alter). As previously mentioned, a histogram of the number of social ties per ego-subject is shown in Fig. 5.1a. Figure 5.5 shows a few examples of geographical distribution of the dwelling places of the ego-subject and his/her social ties. The social tie's residence location was inferred using the method described above. The residence location of

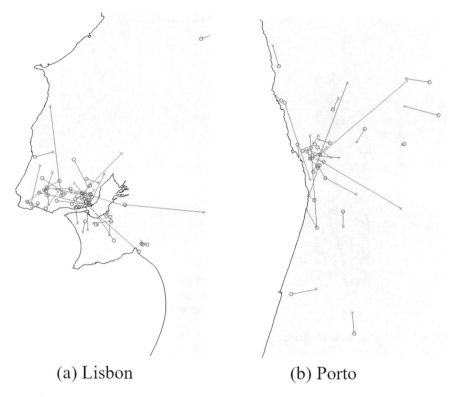

(a) Lisbon   (b) Porto

Fig. 5.3 Urban commuting flows; (**a**) Lisbon and (**b**) Porto. Commuting flows are illustrated with lines, while estimated dwelling and workplace locations are highlighted with red crosses and blue circles, respectively

the ego-subject is marked with a red star, whereas the residence location of the tie is noted with a blue dot. According to [35], social ties are most often geographically grouped near the ego-subject's location. From all ego-subjects, the average distance between the ego-subject and social tie is 39.95 km, with a minimum of 0 km and a maximum of 1701.74 km. Figure 5.6a shows a histogram of distances between the ego-subject's and tie's residence locations. A logarithmic scale was employed to depict the histogram's distribution due to its statistical distribution. There is a peak at 10 km and another peak at 300 km, which is an approximate distance between Lisbon and Porto.

In a social network, there are typically multiple levels of closeness in relation that characterize the strength of a social bond. Blondel et al. [31] looked at various ways for determining a meaningful metric for determining the importance of a relationship between ties. As a result, we used the idea of tie strength introduced by Mark Granovetter in his 1973 milestone paper [79] to deduce the social tie strength. He characterized the strength of a tie as a "combination of duration, emotional intensity, closeness (mutual confiding), and reciprocal services". We took a similar approach to Onnela et al. [69] by using the amount of time spent in communication and

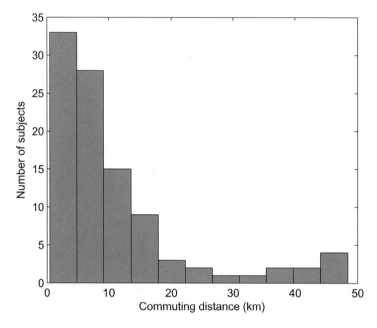

Fig. 5.4 Histogram of commuting distance

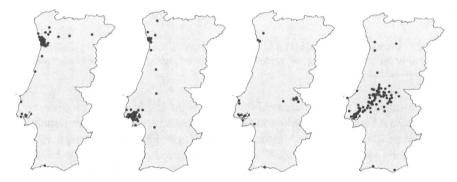

Fig. 5.5 Some examples of the ego-subject's and tie's residence locations. These examples were chosen to demonstrate the socio-geography of certain ego-subjects who live in different parts of the country

reciprocity as proxies. We calculated the tie strength between the ego-subject and a tie based on the overall call duration between them, normalized by the total call duration between the ego-subject and all ties, as given by Eq. (5.1).

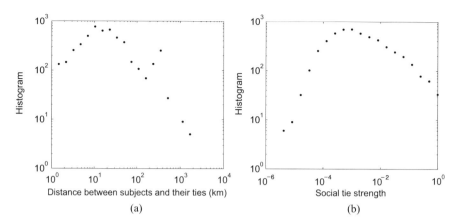

Fig. 5.6 (**a**) Histogram of distances between the ego-subject's and tie's residence locations. (**b**) Histogram of social tie strength values

$$s(i) = \frac{c(i)}{\sum\limits_{i=1}^{N} c(i)}, \tag{5.1}$$

where $s(i)$ is the tie strength between the ego-subject and the ith tie, $c(i)$ is the total call duration between the ego-subject and i^{th} tie, and the denominator is the total call duration between and the ego-subject and all associated ties where N is the total number of associated ties.

As a result, the value of tie strength varies from 0 (lowest strength) to 1 (highest strength). A histogram of social tie strength across all ego-subjects and their ties is shown in Fig. 5.6b. Due to the statistical distribution of the social tie strength values, a logarithmic scale was used, similar to Fig. 5.6a. The average tie strength is 0.018, with 2.1310^{-16} and 0.99 being the minimum and maximum values, respectively.

5.2.4 Transport Mode Inference

Based on the subject's mobile phone usage history, we deduced about the subject's commute mode choice using the estimated residence and workplace locations. We used the Google Maps Directions API [80] as a tool to assist us in making an informed decision about commute mode. We used the Google Maps Direction API to make HTTP requests for waypoints for each subject, using their home and workplace locations as the origin and destination parameters. The two most common modes of transportation today are public transportation and private automobiles. Both transportation and behavioural science studies have looked into the aspects that influence people's decision to use various modes [81–83]. In Portugal, public authorities at both the central and local levels have made recent efforts to promote

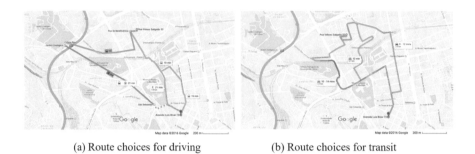

(a) Route choices for driving   (b) Route choices for transit

Fig. 5.7 Routes proposed by the Google Maps Directions API for (**a**) driving and (**b**) transit between a subject's home and workplace locations

the provision of soft modes (such as walking and cycling), but the implementation of a soft mobility network is still in its early stages. As a result, soft modes are rarely used for daily commuting, especially in urban areas [84]. Therefore, in this study, we considered driving and public transit as modes of transportation, and for each subject, two HTTP requests about mode selection were sent. "Driving" and "transit" were utilized as mode parameters in addition to the origin and destination parameters. Routes, or route choices between origin and destination via each transport mode, are included in the answer to each HTTP request (i.e., driving and transit). The amount of route options returned vary depending on the origin and destination locations. The Google Maps Directions API suggested these routes as possible alternatives. Figure 5.7 depicts an example of a route decision between one of the ego-subjects' house and workplace locations. In this example, there are three route options for each mode of transportation.

Our aim was to determine the most likely path followed by the subject based on the traces of the subject's mobile phone activity using these suggested routes. Waypoints (or geographic points) along the route, are included in each return route (recommended route) that we acquired from our HTTP request. Figure 5.8 depicts suggested driving and transit routes (three options for each mode in this example) between a subject's residence and workplace locations, as well as the subject's mobile phone usage history locations, each marked with a green halo circle whose size corresponds to the amount of usage (the number of connections to the cellular network). Driving routes are shown in blue, while transport routes are shown in magenta. Solid red circles indicate the locations of people's homes and workplaces.

The path that is geographically closer to the sites of mobile phone connectivity, i.e., visited areas, is more likely to be the actual route followed by the subject, according to intuition. As an example (in Fig. 5.8), the driving route chosen in the top right corner of the image looks to be the most likely commuting path, as it passes through areas where mobile phones are frequently used. Because the subject has connected to the cell towers along the route collectively (over a year), it is reasonable to assume that this is the route travelled by the subject—and so driving is the mode of choice.

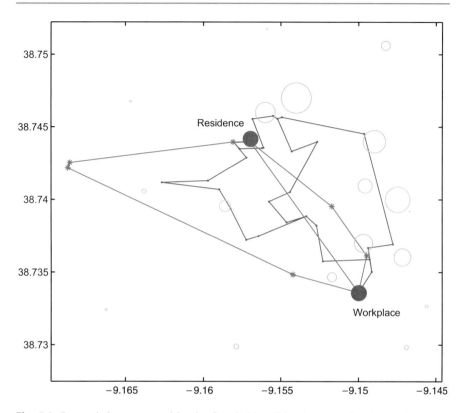

Fig. 5.8 Route choices suggested by the Google Maps Directions API for *driving* (blue lines marked with dots) and *transit* (magenta line marked with asterisks) between a subject's residence and workplace locations. Each location of mobile phone usage history is marked with a circle whose size represents the amount of usage

To ensure fairness in comparisons between different routes in terms of distance to mobile phone usage places, we interpolated the derived routes so that each route contains equally 100 waypoints. The goal was to identify the shortest route between the mobile usage areas and the workplace. The number of data points (waypoints) of 100 for our data interpolation was arbitrarily chosen. Computationally, for each route k (denoted by r_k); $r_k = \{x_k(1), x_k(2), x_k(3), \ldots, x_k(M)\}$, where $x_k(i)$ is the ith waypoint of route k and M is the number of waypoints (100), we computed the distance (m_k) between the M waypoints and a set of all N mobile phone usage locations, as follows.

$$m_k = \frac{1}{M} \sum_{i=1}^{M} c(i), \qquad (5.2)$$

where $c(i)$ is the average haversine distance $(dist(x, y))$ [85] from the waypoint i to all N mobile usage locations, i.e.,

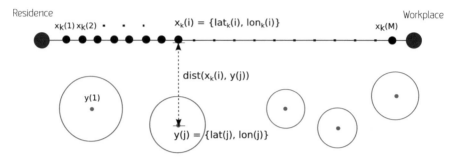

Fig. 5.9 A graphical representation of the distance between the waypoints (black dots) on a journey and the sites where mobile phones are used (green halo circles)

$$c(i) = \frac{1}{N} \sum_{j=1}^{N} dist(x_k(i), y(j)) \cdot w(j) \tag{5.3}$$

Each waypoint $x_k(i)$ consists of geographic coordinates (latitude, longitude); $x_k(i) = \{lat_k(i), lon_k(i)\}$ while each mobile phone usage location $(y(j))$ also consists of geographic coordinates, i.e., $y(j) = \{lat(j), lon(j)\}$ (as shown in Fig. 5.9), therefore $c(i)$ can be calculated as:

$$c(i) = \frac{1}{N} \sum_{j=1}^{N} 2r \cdot \arcsin$$

$$\times \left(\sqrt{\sin^2\left(\frac{lat(j) - lat_k(i)}{2}\right) + \cos(lat_k(i))\cos(lat(j))\sin^2\left(\frac{lon(j) - lon_k(i)}{2}\right)} \right) \cdot w(j), \tag{5.4}$$

where $w(j)$ is a weight function that varies with the level of connectivity at the usage location j as follows:

$$w(j) = \frac{1}{f(j)} \sum_{n=1}^{N} f(n), \tag{5.5}$$

where $f(j)$ is the number of connections (i.e., calls made or received) at the mobile usage location j, so a mobile usage location with a higher connectivity has a lower weight, and vice versa.

For each suggested route, the distance m_k is calculated, and the route with the smallest m_k is chosen as the most likely route taken by the subject, and the mode choice is determined accordingly. With our transport mode inference method, we discovered that driving is the mode of choice for 4500 of the 5405 respondents, while transit is the mode of choice for 905 of them, accounting for 83.26% (driving) and 16.74% (transit). These percentages correspond to actual survey statistics from

Eurostat [86] and ECORYS Transport of Portugal [87], which show 85% for driving and 15% for transport. For our ego-subjects, driving is the choice of 67 subjects, while transit is the choice of 33 subjects.

5.3   Results

5.3.1   Commute Mode Choices of Social Ties

The inferred commute transport modes and social ties allowed us to dig further into the ego-subjects' social ties mode choices. We divided the ego-subjects into two groups based on their preferred mode of transportation: driving and public transportation. We examined the mode choices of the ego-subjects' ties for each ego-group subject's because our original objective was to investigate the typical distributions of social ties' mode choices that were considered to impact the choices of the ego-subjects.

We looked at the number of driving and transit ties for each ego-subject and estimated the percentage of each tie's mode choice. A histogram of the fraction of each mode option of the driving ego-subjects' social ties is shown in Fig. 5.10a. For the driving ties, the average and standard deviation were calculated to be 0.855 and 0.828, respectively, and for the transit ties, 0.145 and 0.828. It can be shown that driving is chosen by a substantially higher percentage of social ties, which may have influenced the mode choice decision for the driving ego-subjects.

By looking at the mode choices of the transit ego-subjects' social ties, we discovered that the fraction of social ties who choose driving is still higher than the fraction of social ties who choose transit, as shown in Fig. 5.10b. For driving ties, the average and standard deviation are (0.791, 0.852), whereas for transit ties, they are (0.209, 0.852).

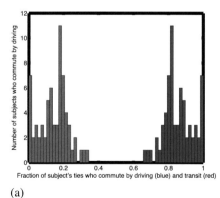

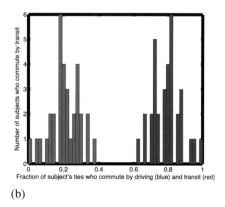

(a)   (b)

Fig. 5.10 (**a**) Histogram of the fraction of each mode choice of the driving ego-subjects' social ties; driving (blue bars), transit (red bars). (**b**) Histogram of the fraction of the transit ego-subjects' social ties of each mode choice; driving (blue bars), transit (red bars)

As driving is the most popular form of transportation, its fractions appear to outnumber the fractions of transit ties in both ego-subjects' mode choice groups. Despite this, there is a modest shift in these fractions, with the average of the transit ties' percentage (0.209) slightly greater than the driving ego-subjects' transit ties' fraction (0.145), and a downshift in the driving ties' fractions for transit ego-subjects. This minor evidence may have implied that the mode choices made by the members of the ego-subject social network can influence the ego-subject mode's choice as well.

We re-generated the result with a random reference system that randomly shuffles the mode selections among subjects in the dataset to guarantee that this finding is not completely due to the unbalanced distribution of vehicle and transit users. To begin, we rearranged people's driving mode preferences and recalculated the average fractions of driving and transit social links for each set of driving and transit ego-subjects. This experiment was repeated ten times. Second, we re-measured the findings after rearranging the transit mode options. Since there are overlaps, the differences between the average fractions of driving and transit ties are less than the results observed in Fig. 5.10, which are (0.855, 0.145) for driving subjects and (0.791, 0.209) for transit subjects. The average difference between the fractions of driving and transit ties from the random reference system is 0.587 for driving and 0.527 for transit subjects, respectively, whereas the results in Fig. 5.10 have differences of 0.710 and 0.582 for driving and transit subjects, respectively—which are overall 12.3 and 5.5% greater than the random references. This suggests that the result in Fig. 5.10 is influenced by social factors as well as the unbalanced distribution of car and transit users.

There could be several reasons why the ego-subjects' mode choices mirror the mode choices of the ego-subjects' ties for each ego-subject's group. Other members of the ego-social subject's network, according to Pike [59], can influence the ego-subject. Alternatively, it is possible that all members of the ego-social subject's network have a comparable decision context, such as similar costs associated with each mode, resulting in the ego-subject social network choosing the same mode. It's also feasible that members of the subject's social network are all prone to making the same decision due to shared transportation attitudes, and this early discovery has prompted us to explore it further.

5.3.2   Social Distance

The degree of social closeness or relationship is determined by the strength of ties. People who are socially close to us and whose social circles closely coincide with our own are considered strong ties. They are usually folks we trust and who share a number of common interests with us. Weak ties, on the other hand, represent mere acquaintances. Various behaviors, such as information receipt [88], mobility [60], and word-of-mouth referral [89], might be influenced differently depending on the degree of social ties.

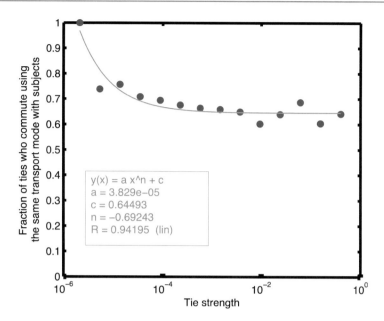

Fig. 5.11 Fraction of social ties with individual ego-subjects who share a common commute mode choice, spanning a range of tie strength values

Along the same lines of behavior analysis, we were curious about the relationship between tie strength and the mode of transportation choice decision. As a result, we expanded our exploratory analysis into social ties' commute mode preferences to look at the strength of the ties. We looked at the percentage of social links that have the same mode choice as the individual ego-subjects across a variety of tie strength values (calculated using Eq. (5.1)). Interestingly, the percentage of ties that share a similar mode choice falls as the strength of the tie grows. This relationship can be fitted with a power law equation $y(x) = 3.829 \times 10^{-5}x^{-0.69243} + 0.64493$ with $r = 0.941$, as shown in Fig. 5.11. As a bigger proportion of them share a common mode choice with the ego-subjects, the result may suggest that weaker relationships have a greater influence on the ego-subject's mode choice. To ensure that the observed outcome is not solely attributable to the overwhelming fraction of weak ties, we re-created it using a network with the same number of connections, degree, and overall fraction of weak ties as the original, but with people's transportation modes reshuffled. For a total of ten trials, we repeated the experiment. When compared to the result in Fig. 5.11, which is more structured and can be nicely fitted by an equation, the results obtained from this setup are rather random.

We explored two different sets of ego-subjects: driving and transit. The driving ego-subjects (shown in Fig. 5.12a) have a very similar outcome, while the transit ego-subjects (shown in Fig. 5.12b) do not. Two unique tendencies are seen for driving ego-subjects, which is interesting. When the tie strength is smaller than the average tie strength, the percentage of driving ties reduces ($y(x) = 0.0015311x^{-0.37327} + 0.81391$ with $r = 0.943$). When the tie strength is greater

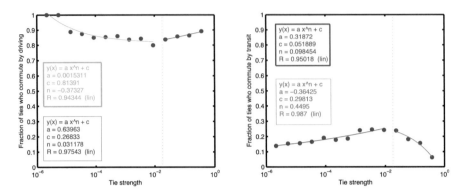

Fig. 5.12 (a) Fraction of social ties sharing a common commute mode choice with the driving ego-subjects, across a range of tie strength values. Red vertical dash line indicates the average tie strength value (b) fraction of social ties sharing a common commute mode choice with the transit ego-subjects, across a range of tie strength values. The average tie strength value is indicated by the red vertical dash line

than the average, however, the proportion of driving ties increases ($y(x) = 0.63963x^{-0.031178} + 0.26833$ with $r = 0.975$).

Strong ties and weak ties [60] have been determined using the average tie strength, with those with a tie strength value more than the average being categorized as strong ties and those with a tie strength value less than the average being classified as weak ties. As a result of the data in Fig. 5.12a, weak ties become less influential as tie strength increases, while strong ties become more influential for the driving ego-subjects. Note that being influential in this context indicates that the ego-subject is more likely to have a common decision, as seen by the fraction of social ties' mode choices observed.

Individuals with tenuous or even random relationships—i.e., weak ties—are often impacted by those with whom they have tenuous or even random relationships, as stated by Granovetter [79]. Our findings support Goldenberg et al. [90]'s investigation of the impact of strong and weak ties on the acceptance of a new product, which found that the influence of weak ties is as powerful as strong ties, and their effect approximates or exceeds that of strong ties at all stages of the product life cycle. They also discovered that when personal networks are small, weak ties have a greater impact on information dissemination than strong ties, which is likely the case for our result in Fig. 5.12a. Our result thus adds to the literature from the perspective of the influence of social ties on the choice of commuting by a car.

The transit ego-subjects, on the other hand, show the opposite pattern (Fig. 5.12b). Weak ties have a stronger influence as tie strength grows ($y(x) = 0.31872x^{-0.098454} + 0.051889$ with $r = 0.950$), whereas strong ties have a less influence ($y(x) = -0.36425x^{-0.4495} + 0.29813$ with $r = 0.987$). As a result, Fig. 5.12 suggests that strong ties are more important to determine if driving is a person's preferred mode of transportation, whereas weak ties are more important to determine if public transportation is a person's preferred mode of transportation.

People who take public transportation have a higher possibility of forming homophilic weak links since they share space and spend time together. On the other hand, driver-commuters are either traveling alone or with persons who are apparently strong links. This finding adds to the findings of de Kleijn [91], which found that stronger ties had a bigger influence on public transportation choice than weaker ties, and it extends the observation by noting that the influence of strong ties decreases as tie strength grows.

5.3.3 Physical Distance

Following our investigation into the impact of social distance on mode choices, we wanted to see how geographical distance influences mode choice decisions. We wanted to see if physical distance to social ties is also an influential element in mode choice decision, as we've already seen that distance in social relationships is an influential component. Is it true that friends who live (or work) close by have more influence than friends who live (or work) further away? As different geographical areas may be organized with distinct physical arrangements, the question of landscape and transportation infrastructures may influence the mode choice decision.

As a result, we looked at the percentage of ties that have a common mode choice with the individual ego-subjects over a variety of geographical distances. The locations of one's home and workplace were taken into consideration. We noticed that when the distance between ties grows, the percentage of ties who have a common mode choice decreases (as shown in Fig. 5.13). The drop is approximately 10% from 0 to 600 km in all cases. The fitted curve equations and their corresponding correlation coefficient r are: $y(x) = 1.2549(x + 86.058)^{-0.13877}$ with $r = 0.924$ (between the ego-subject's and tie's residence locations), $y(x) = 0.9135$ $(x + 42.312)^{-0.07842}$ with $r = 0.843$ (between the ego-subject's residence and tie's workplace locations), $y(x) = 0.92355(x + 55.013)^{-0.080695}$ with $r = 0.832$ (between the ego-subject's workplace and tie's residence locations), and $y(x) = 10.02$ $(x + 649.35)^{-0.4196}$ with $r = 0.888$ (between ego-subject's and tie's workplace locations).

The findings indicate that social links that are geographically nearby have a greater influence on commute mode choice than those that are farther away. Landscape and transportation infrastructures, for example, may play a role in mode choice since they constrain spatial layout and consequently transportation in the area. As a result, folks who live in close proximity have more similar transportation options.

Previous research looked at the interaction between mobility patterns and social relationships from a different perspective. Their main goal was to determine the links between tie strength and mobility similarity, and they discovered that mobility similarity can be utilized to categorize social interactions [62, 63]. Physical distance to social ties, from working/living locally to sharing/attending groups of comparable sites, can be an influencing component in mode choice decision, according to our research.

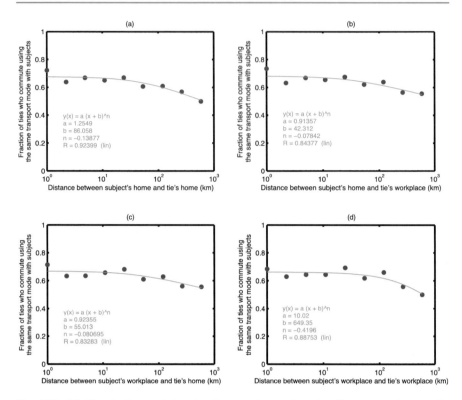

Fig. 5.13 Likelihood of a social tie using the same transport mode with the ego-subject as the distance between the ego-subject and the tie's locations varies; (**a**) subject's residence and tie's residence locations, (**b**) ego-subject's residence and tie's workplace locations, (**c**) ego-subject's workplace and tie's residence locations, and (**d**) ego-subject's workplace and tie's workplace locations

We kept asking the question about physical space or spatial organization, specifically in relation to public transit accessibility — that is, how public transit infrastructure influences the decision to use public transportation. The accessibility of public transportation is critical for assessing current services and forecasting travel demand. One of the accessibility measures is the access distance [64]. So we measured the distance to the nearest public transportation stop for each of our 4405 subjects. We used the Google Places API with the "Type" parameter set to 'transit station' via the Nearby Place Search [65]. The percentage of subjects that utilize public transportation was next analyzed, as the distance to the nearest public transit station ranged from 100 m to 10 km. As expected, the percentage of transit users reduces as the access distance increases (see Fig. 5.14). This relationship can also be described by a fitted power law equation: $y(x) = 1.2526 \times 10^6 (x + 8454.7)^{-1.7181}$ with $r = 0.888$. As a result, the distance to public transportation is one of the deciding variables in commuting mode selection.

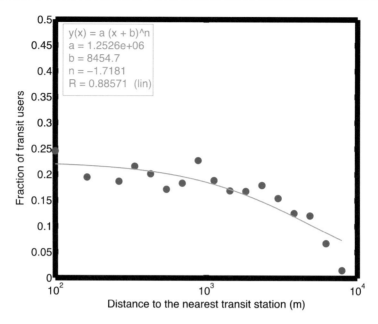

Fig. 5.14 As the distance to the nearest transit station increases, the likelihood of using public transportation drops

Accessibility is defined as the ease with which people can get to public transportation terminals in this study. Other characteristics of public transportation use, such as out-of-vehicle and in-vehicle-times, speed, directness of travel, and number of transfers for specific origin-destination connections, will be explored in future studies. Papaioannou and Martinez [92] stressed the importance of accessibility in people's mode choice decisions in a study done in Lisbon. This is not always the case; in large cities where citizens have access to a number of transportation options, distance is less of an issue, and travelers prefer to use modes that give better service rather than those that are closer.

5.3.4   Ego-Network Effect

We are back to the topic of how our social network affects our mode choice decision. Do we have a tendency of having the same choice with people in our social network? Is a person more inclined to take public transportation if the majority of the people in their social network do? To investigate the ego-network impact, we looked at the percentage of social ties in each ego-subject social network that had a commute mode preference.

We looked at the percentage of social ties that had the same commute method as the main subject. The outcome (shown in Fig. 5.15) reveals that when the percentage of ties who use the same commuting mode choice increases within his or her social

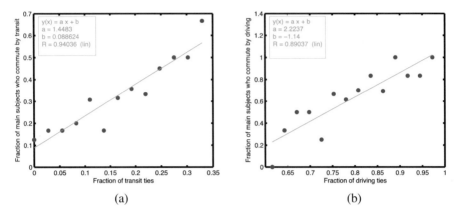

Fig. 5.15 Likelihood of choosing (**a**) transit or (**b**) driving as a commute mode choice increases as the portion of social ties using the same mode choice increases

network, the subject is more likely to use that mode with his or her ties. This is true for both modes of transit (a) and driving (b). For transit, the trend can be described by a fitted linear equation $y(x) = 1.4483x + 0.088624$ with $r = 0.940$, while for the driving, the fitted linear equation is $y(x) = 2.2437x - 1.14$ with $r = 0.89$.

As demonstrated in this study, persons who are already connected inside a personal network might impact one another's behavioral features, such as commuting mode choice. Our data do not allow us to go any further in explaining the psychological elements that influence individual decision-making, but psychological research has found that social expectations influence intention and habits to drive or take public transportation [93]. According to our findings, social influence is most effective when it originates in one's interpersonal relations.

5.4   Conclusion

The advancement of information and telecommunication technologies, such as personal mobile phones, has opened up new exciting opportunities for behavior studies, one of which is opportunistic behavioral sensing, which uses a cellular network to incorporate humans as part of the sensing infrastructure. In a case study discussed in this chapter, we were able to infer transportation mode choices of 4405 mobile phone users who were classified as subjects in our study, as well as analyze the social influence on commute mode choice decision, by analyzing longitudinal mobile phone data that included both location and communication logs. Our findings reveal that strong ties are more crucial in determining whether a person prefers to drive or not. Weak ties, on the other hand, are more relevant in determining whether or not a person prefers to use public transportation. These findings appear to support the threshold model of social influence [7]. However, it also suggests that the impact of others can be classified based on their relation with ego. Not only are not all people equally impacted by others [94], but different links can have varying degrees of influence on mode choice.

Furthermore, we've discovered that social ties that are geographically nearby have a greater impact on commuting mode choice than those that are farther away. Our findings also reveal that the distance to public transportation is one of the deciding variables in commuting mode choice, as the percentage of transit users drops as the distance to public transportation increases. Moreover, we've also discovered that the likelihood of choosing either transportation or driving as a commuting option improves as the percentage of social ties who use that mode grows. As a result, it is clear that social networks have an impact on transportation mode selection.

Clearly, there are certain limitations in the observations made in this study. The first of these is the dataset's location traces' discontinuous characteristic. We were only able to identify a subset of all the locations visited in a day since individuals were only found when connections with the cellular network were established. However, to a great extent, the longitudinal character of our data and the aggregate of mobility patterns compensate for this. The coarse spatial resolution of the location traces, which is dictated by the granularity of cellular tower coverage, is the second limitation. Traditional surveys could reach much higher spatial resolution, but they are only carried out for small samples of the entire population and for relatively short periods of time in practice. Another limitation is the lack of individual ground-truth information for individual modes of transportation, which limits our validation to an aggregate level (national level). An individual-level validation is thus worth investigated in the future as a component of inquiry into possible ground-truth data collection methods. Possible future research also includes a creation of a map matching technique that can be used to better connect waypoints (legs information) to the actual road network. Furthermore, other important places besides home and work, such as a grocery store, restaurant, and so on, may be investigated for a more complete understanding of travel behavior, particularly mode choice behavior, which can vary significantly depending on the type of these contextual places, as evidenced by service usage patterns [27, 28]. Moreover, various essential aspects such as age, socioeconomic position, geographical limits, social influence, environmental awareness, commuting duration, and so on might impact transportation mode choice decisions. One of these many powerful elements is social influence, which is the subject of this study. Although social influence is not the most important factor of travel behavior, it does play a role, and its impact has been overlooked when modeling the mode choice decision process. As a result, this study presents an observational view of travel behavior in the context of social networks.

The study discussed in this chapter adds to the social context of transportation behavior research as it introduces new framework to analyze the social influence on transportation mode choice behavior using mobile phone network data, which is based on a static, ego-centered social network. With time-evolving information such as transportation mode adoption and network evolution, an investigation of additional insights into societal influence on mode adoption will be possible in the future.

Future investigation may also consider incorporating additional data sources, such as socioeconomic status in various locations.

References

1. Litman T. The new transportation planning paradigm. Inst Transp Eng J. 2013;83(6):20–4.
2. Goodwin P. Three views on 'peak car'. World Transp Policy Pract. 2011;17(4):8–17.
3. Schwanen T, Banister D, Anable J. Rethinking habits and their role in behaviour change: the case of low-carbon mobility. J Transp Geogr. 2012;24:522–32.
4. Domencich T, McFadden D. Statistical estimation of choice probability functions. 1975.
5. Ortúzar J, Willusen L. Modeling transport. 4th ed; 2011.
6. Axhausen KW. Social networks and travel: some hypotheses. 2003;197.
7. Granovetter M. Threshold models of collective behavior. Am J Sociol. 1978;83(6):1420–43.
8. McPherson M, Smith-Lovin L, Cook JM. Birds of a feather: homophily in social networks. Annu Rev Sociol. 2001;27(1):415–44.
9. Newman MEJ. Mixing patterns in networks. Phys Rev E. 2003;67(2):026126.
10. Kowald M, Arentze TA, Axhausen KW. Individuals' spatial social network choice: model-based analysis of leisure-contact selection. Environ Plan B Plan Des. 2015;42(5):857–69.
11. Bar-Gera H. Evaluation of a cellular phone-based system for measurements of traffic speeds and travel times: a case study from Israel. Transp Res Part C Emerg Technol. 2007.
12. Demissie MG, de Almeida Correia GH, Bento C. Intelligent road traffic status detection system through cellular networks handover information: an exploratory study. Transp Res Part C Emerg Technol. 2013.
13. Herrera JC, Work DB, Herring R, Ban X, Jacobson Q, Bayen AM. Evaluation of traffic data obtained via GPS-enabled mobile phones: The Mobile Century field experiment. Transp Res Part C Emerg Technol. 2010;18(4).
14. Liu HX, Danczyk A, Brewer R, Starr R. Evaluation of cell phone traffic data in Minnesota. Transp Res Rec J Transp Res Board. 2008;2086(1):1–7.
15. Caceres N, Wideberg JP, Benitez FG. Deriving origin–destination data from a mobile phone network. IET Intell Transp Syst. 2007;1(1):15–26.
16. Demissie MG, Antunes F, Bento C, Phithakkitnukoon S, Sukhvibul T. Inferring origin-destination flows using mobile phone data: a case study of Senegal. In: 2016 13th international conference on electrical engineering/electronics, computer, telecommunications and information technology (ECTI-CON); 2016. p. 1–6.
17. Iqbal MS, Choudhury CF, Wang P, González MC. Development of origin–destination matrices using mobile phone call data. Transp Res Part C Emerg Technol. 2014;40:6374.
18. Pan C, Lu J, Di S, Ran B. Cellular-based data-extracting method for trip distribution. Transp Res Rec J Transp Res Board. 2006;1945(1):33–9.
19. White J, Wells I. Extracting origin destination information from mobile phone data. In: Road transport information and control, vol. 486; 2002. p. 30–4.
20. Alexander L, Jiang S, Murga M, González MC. Origin–destination trips by purpose and time of day inferred from mobile phone data. Transp Res Part C Emerg Technol. 2015;58:240–50.
21. Çolak S, Alexander LP, Alvim BG, Mehndiratta SR, González MC. Analyzing cell phone location data for urban travel. Transp Res Rec J Transp Res Board. 2015;2526(1):1–17.
22. Demissie MG, Phithakkitnukoon S, Sukhvibul T, Antunes F, Gomes R, Bento C. Inferring passenger travel demand to improve urban mobility in developing countries using cell phone data: a case study of Senegal. IEEE Trans Intell Transp Syst. 2016;17(9):2466–78.
23. Demissie MG, Correia G, Bento C. Analysis of the pattern and intensity of urban activities through aggregate cellphone usage. Transp A Transp Sci. 2015.
24. Soto V, Frias-Martinez E. Robust land use characterization of urban landscapes using cell phone data. 2011.

25. Toole JL, Ulm M, González MC, Bauer D. Inferring land use from mobile phone activity. In: Proceedings of the ACM SIGKDD international conference on knowledge discovery and data mining; 2012. p. 1–8.
26. Trestian I, Ranjan S, Kuzmanovic A, Nucci A. Measuring serendipity. In: Proceedings of the 9th ACM SIGCOMM conference on internet measurement conference – IMC '09; 2009. p. 267.
27. Jo H-H, Karsai M, Karikoski J, Kaski K. Spatiotemporal correlations of handset-based service usages. EPJ Data Sci. 2012;1(1):10.
28. Karikoski J, Soikkeli T. Contextual usage patterns in smartphone communication services. Pers Ubiquit Comput. 2013;17(3):491–502.
29. Steenbruggen J, Borzacchiello MT, Nijkamp P, Scholten H. Mobile phone data from GSM networks for traffic parameter and urban spatial pattern assessment: a review of applications and opportunities. GeoJournal. 2013;78(2):223–43.
30. Calabrese F, Ferrari L, Blondel VD. Urban sensing using mobile phone network data: a survey of research. ACM Comput Surv. 2014;47(2):25.
31. Blondel VD, Decuyper A, Krings G. A survey of results on mobile phone datasets analysis. EPJ Data Sci. 2015.
32. Mutz DC. Impersonal influence: effects of representations of public opinion on political attitudes. Polit Behav. 1992;14(2):89–122.
33. Helbing D, Farkas I, Vicsek T. Simulating dynamical features of escape panic. Nature. 2000;407(6803):487–90.
34. Hirshleifer D, Hong Teoh S. Herd behaviour and cascading in capital markets: a review and synthesis. Eur Financ Manag. 2003;9(1):25–66.
35. Krumme C, Cebrian M, Pickard G, Pentland S. Quantifying social influence in an online cultural market. PLoS One. 2012;7(5):e33785.
36. Schweitzer F, Mach R. The epidemics of donations: logistic growth and power-laws. PLoS One. 2008;3(1):e1458.
37. Sridhar S, Srinivasan R. Social influence effects in online product ratings. J Mark. 2012;76(5): 70–88.
38. Mavrodiev P, Tessone CJ, Schweitzer F. Quantifying the effects of social influence. Sci Rep. 2013;3(1):1360.
39. Prechter RR. Unconscious herding behavior as the psychological basis of financial market trends and patterns. J Psychol Financ Mark. 2001;2(3):120–5.
40. Wenzel M. Misperceptions of social norms about tax compliance: from theory to intervention. J Econ Psychol. 2005;26(6):862–83.
41. Leahey TM, Kumar R, Weinberg BM, Wing RR. Teammates and social influence affect weight loss outcomes in a team-based weight loss competition. Obesity. 2012;20(7):1413–8.
42. Salomon I. Telecommunications and travel – substitution or modified mobility? J Transp Econ Policy. 1985;19(3):219–35.
43. Hägerstraand T. What about people in regional science? Pap Reg Sci. 1970;24(1):7–24.
44. Janelle DG, Goodchild MF, Klinkenberg B. Space-time diaries and travel characteristics for different levels of respondent aggregation. Environ Plan A Econ Sp. 1988;20(7):891–906.
45. Bo L. Path in space-time environments: a time-geographic study of the movement possibilities of individuals. Environ Plan. 1976;9(8):961–72.
46. Pred A. Of paths and projects: individual behavior and its societal context. New York: Methuen; 1981.
47. Harvey AS, Taylor ME. Activity settings and travel behaviour: a social contact perspective. Transportation (Amst). 2000;27(1):53–73.
48. Arentze T, Timmermans H. Social networks, social interactions, and activity-travel behavior: a framework for microsimulation. Environ Plan B Plan Des. 2008;35(6):1012–27.
49. Gordon P, Kumar A, Richardson HW. Gender differences in metropolitan travel behaviour. Reg Stud. 1989;23(6):488–510.
50. Hanson S, Hanson P. The impact of married women's employment on household travel patterns: a Swedish example. Transportation (Amst). 1981;10(2):165–83.

51. Hanson S, Hanson P. The travel-activity patterns of urban residents: dimensions and relationships to sociodemographic characteristics. Econ Geogr. 1981;57(4):332–47.
52. Pas EI. The effect of selected sociodemographic characteristics on daily travel-activity behavior. Environ Plan A Econ Sp. 1984;16(5):571–81.
53. Lu X, Pas EI. Socio-demographics, activity participation and travel behavior. Transp Res Part A Policy Pract. 1999;33(1):1–18.
54. Carrasco JA, Hogan B, Wellman B, Miller EJ. Collecting social network data to study social activity-travel behavior: an egocentric approach. Environ Plan B Plan Des. 2008;35(6):961–80.
55. Anderson DA, Ben-Akiva M, Lerman SR. Discrete choice analysis: theory and applications to travel demand. J Bus Econ Stat. 1988;6(2):286.
56. Gliebe JP, Koppelman FS. A model of joint activity participation between household members. Transportation (Amst). 2002;29(1):49–72.
57. Scott DM, Kanaroglou PS. An activity-episode generation model that captures interactions between household heads: development and empirical analysis. Transp Res Part B Methodol. 2002;36(10):875–96.
58. Páez A, Scott DM. Social influence on travel behavior: a simulation example of the decision to telecommute. Environ Plan A Econ Sp. 2007;39(3):647–65.
59. Scott DM, Dam I, Páez A, Wilton RD. Investigating the effects of social influence on the choice to telework. Environ Plan A Econ Sp. 2012;44(5):1016–31.
60. Phithakkitnukoon S, Smoreda Z, Olivier P. Socio-geography of human mobility: a study using longitudinal mobile phone data. PLoS One. 2012;7(6):e39253.
61. Song C, Qu Z, Blumm N, Barabási AL. Limits of predictability in human mobility. Science. 2010;327(5968):1018–21.
62. González MC, Hidalgo CA, Barabási AL. Understanding individual human mobility patterns. Nature. 2008;453:779–82.
63. Song C, Koren T, Wang P, Barabási AL. Modelling the scaling properties of human mobility. Nat Phys. 2010;6:818–23.
64. Calabrese F, Di Lorenzo G, Ratti C. Human mobility prediction based on individual and collective geographical preferences. In: 13th international IEEE conference on intelligent transportation systems; 2010. p. 312–7.
65. Becker R, et al. Human mobility characterization from cellular network data. Commun ACM. 2013;56(1):74–82.
66. Wesolowski A, et al. Quantifying the impact of human mobility on malaria. Science. 2012;338 (6104):267–70.
67. Wang P, González MC, Hidalgo CA, Barabási A-L. Understanding the spreading patterns of mobile phone viruses. Science. 2009;324(5930):1071–6.
68. Hidalgo CA, Rodriguez-Sickert C. The dynamics of a mobile phone network. Phys A Stat Mech its Appl. 2008;387(12):3017–24.
69. Onnela J-P, et al. Structure and tie strengths in mobile communication networks. Proc Natl Acad Sci. 2006;104(18):7332–6.
70. Phithakkitnukoon S, Dantu R. Mobile social group sizes and scaling ratio. AI Soc. 2011;26(1): 71–85.
71. Phithakkitnukoon S, Calabrese F, Smoreda Z, Ratti C. Out of sight out of mind – how our mobile social network changes during migration. In: IEEE international conference on privacy, security, risk and trust and IEEE international conference on social computing (PASSAT/ SocialCom 2011); 2011. p. 515–20.
72. Eagle N, de Montjoye Y-A, Bettencourt LMA. Community computing: comparisons between rural and urban societies using mobile phone data. In: 2009 international conference on computational science and engineering; 2009. p. 144–50.
73. Eagle N, Macy M, Claxton R. Network diversity and economic development. Science. 2010;328(5981):1029–31.
74. Phithakkitnukoon S, Leong TW, Smoreda Z, Olivier P. Weather effects on mobile social interactions: a case study of mobile phone users in Lisbon, Portugal. PLoS One. 2012;7(10).

75. Onnela J-P, Arbesman S, González MC, Barabási A-L, Christakis NA. Geographic constraints on social network groups. PLoS One. 2011;6(4):e16939.
76. Krings G, Calabrese F, Ratti C, Blondel VD. Scaling behaviors in the communication network between cities. In: 2009 international conference on computational science and engineering; 2009. p. 936–9.
77. Calabrese F, Smoreda Z, Blondel VD, Ratti C. Interplay between telecommunications and face-to-face interactions: a study using mobile phone data. PLoS One. 2011;6(7):e20814.
78. De Domenico M, Lima A, Musolesi M. Interdependence and predictability of human mobility and social interactions. Pervasive Mob Comput. 2013;9(6):798–807.
79. Granovetter M. The strength of weak ties. Am J Sociol. 1973;78(1):1360–80.
80. Google. Google Maps Directions API. 2016.
81. Chee WL, Fernandez JL. Factors that influence the choice of mode of transport in Penang: a preliminary analysis. Procedia Soc Behav Sci. 2013;91(10):120–7.
82. Beirão G, Sarsfield Cabral JA. Understanding attitudes towards public transport and private car: a qualitative study. Transp Policy. 2007;14(6):478–89.
83. Anwar AM. Paradox between public transport and private car as a modal choice in policy formulation. J Bangladesh Inst Planners. 1970;2:71–7.
84. Viegas F. Critérios para a Implementação de Redes de Mobilidade Suave em Portugal Um Caso de Estudo no Município de Lagoa. Lisbon: Universidade Técnica de Lisboa Instituto Superior Técnico; 2008.
85. Sinnott R. Virtues of the Haversine. Sky Telesc. 1984;68(2):159.
86. Eurostat. Instituto Nacional De Estatistica (Statistics Portugal). Modal split of passenger transport. 2011.
87. ECORYS Transport. ECORYS transport. In: Study on strategic evaluation on transport investment priorities under structural and cohesion funds for the programming period 2007–2013, no 2005.CE.16.0.AT.014; 2006.
88. Lauwerijssen P. Tie strength and the influence of perception: obtaining diverse or relevant information. Tilburg University; 2011.
89. Brown JJ, Reingen PH. Social ties and word-of-mouth referral behavior. J Consum Res. 1987;14(3):350–62.
90. Goldenberg J, Libai B, Muller E. Talk of the network: a complex systems look at the underlying process of word-of-mouth. Mark Lett. 2001;12(3):211–23.
91. de Kleijn S. The influences of an individual's social network on the choice of travelling by public transport. Utrecht University; 2015.
92. Papaioannou D, Martinez LM. The role of accessibility and connectivity in mode choice. A structural equation modeling approach. Transp Res Procedia. 2015;10:831–9.
93. Donald IJ, Cooper SR, Conchie SM. An extended theory of planned behaviour model of the psychological factors affecting commuters' transport mode use. J Environ Psychol. 2014;40: 39–48.
94. Watts DJ, Dodds P. Threshold models of social influence. Oxford University Press; 2017.

Inferring Route Choice Using Mobile Phone CDR Data

6

Abstract

Telecom operators acquire communication logs of our mobile phone usage activity for billing purposes. These communication records, also known as CDR, have proven to be an important data source for a human behavioral investigation. This chapter describes a framework for collecting data on route choice behavior based on crowdsourcing approach by using CDR data to infer individual commuting trip route choices. In this chapter, we discuss methods for inferring route choice based on a calendar year of CDR data obtained from mobile phone users in Portugal. Interpolation of route waypoints, shortest distance between a route choice and mobile usage locations, and Voronoi cells that assign a route option to coverage zones are the main approaches. These strategies are explored in combination with noise filtering utilizing DBSCAN (Density-Based Spatial Clustering of Applications with Noise) and commuting radius. In comparison to costly and time-consuming traditional travel surveys, the methodology and results discussed in this chapter are valuable for transportation modelling as they give a fresh, viable, and economical means of acquiring route choice data. Moreover, a route choice inference based on CDR data at this level of detail, i.e., street level, has rarely been studied. This chapter reflects the ideas, motivation, and thinking process of the route choice and commuting trip inference in our original work by Sakamanee et al. (*J., 2020, Int., 6:306, Geo-Information, ISPRS*) and Jundee et al. ((UbiComp/ISWC, of, on, ACM, and, the, 2018, 2018), Adjunct, Computers, Computing, Conference, International, Joint, Pervasive, Proceedings, Symposium, Ubiquitous, Wearable).

Keywords

Call detail records · Route choice inference · Commuting trip · Route waypoints · DBSCAN · Commuting radius

S. Phithakkitnukoon, *Urban Informatics Using Mobile Network Data*,
https://doi.org/10.1007/978-981-19-6714-6_6

6.1   Motivation and State of the Art

Commuting is the most common and recurring travel between home and workplace. This accounts for the bulk of journeys taken by individuals. As a result, commuting is believed to be one of the key causes of traffic congestion [1]. Understanding commute patterns is very crucial and beneficial to urban planning and traffic management. Commuting patterns and habits have been studied in the fields of human geography, transportation, and urban studies due to their nature. Nonetheless, due to the recent availability of large-scale electronic datasets from which different models can be examined and developed to better describe characteristics of human mobility at various spatial scales, commuting and mobility patterns are becoming a more appealing research problem for scholars from other disciplines such as physics, statistics, and data science. Commuting and mobility patterns are becoming a more attractive research problem for scholars from other disciplines.

Human mobility as seen in the form of collective trips is studied in the domain of travel demand modeling, for which the sequential four-step model [2] has traditionally been used for transportation forecasts, such as estimating the number of vehicles on a planned road, ridership on a railway line, and bus passengers at the airport. Trip generation, trip distribution, mode choice, and route assignment are the four steps of the model, with each step meant to represent the amounts, locations, travel modes, and route choices of generated trips.

The collecting of traffic data linked to travel behavior—for example, traffic count, number of trips taken from/to a specific location, and start/end times of journeys—is the first step in travel demand modeling. Traditionally, traffic data has been gathered through individual travel habit surveys. Most surveys collect information about the individual, their household, and a journal of their travels on a given day. The most common methods for a travel survey include traffic counts, roadside interviews, and questionnaires, which are all expensive and time-consuming [3]. Thus, comprehensive travel surveys are only done once every 10 years. The collected data may be outdated due to the significant gap between surveys, despite the fact that it includes detailed mobility information. Its high cost also prevents travel surveys from being conducted inside specific study zones, resulting in skewed data. Furthermore, the information gathered frequently relies on survey participants recalling details about previous journeys, which can be misleading due to inaccurate responses to travel survey forms.

Sensors such as GPS tracking units are increasingly being employed for travel surveys. This is as a result of recent improvements in location-aware technologies [4]. However, due to privacy concerns and restrictions, such as the EU general data protection legislation (GDPR) [5], gathering data on such a huge scale is tough and challenging. Recent attempts have yielded data limited to specific types of tracked individuals, such as college students [6], city bikers [7], and service clients [8]. Nonetheless, privacy considerations keep this type of precise mobility data from being widely available and used for travel demand modeling.

Opportunistic sensing data derived from a variety of sources has recently emerged as a promising alternative for gaining insights into the spatial distribution of human mobility. Opportunistic data is information that was acquired for one purpose but now has the potential to be used for another. Mobile phone network data, also known as CDR (call detail records), is a type of opportunistic sensing data that is collected for the purpose of customer billing but can also be used for human mobility research. A CDR is a record of a mobile phone user's cellular network connectivity. Each time the mobile phone user connects to a serviced cellular network by receiving or making a call or using the internet, the communication information is recorded—i.e., call duration, timestamp, caller's and receiver's identifications, and location of the connected cellular tower. Collectively, with these individual location footprints, CDR has emerged as a useful data source in human mobility studies [9, 10]. Though the CDR data is not as detailed as GPS tracking data, it is worth noting that there is still a privacy concern even if they are anonymized when analyzed rigorously with additional outside information [11].

In the context of the four-step model, the CDR data has been used in each of the sequential steps. In the trip generation step, it has been used for—among other things—inference of trip volume and spatial distribution for estimating commuting tip generation rates [12], calibration of a hybrid trip generation model [13], and estimation of zonal travel demand [14]. It has been used in several studies in the trip distribution step, such as origin–destination (O-D) matrix's construction [15–17], evaluation [18], and modeling [19–21]. In the mode choice step, the CDR data has been used for inferring commuting mode choice based on distance measures between visited cell towers and route choices [22], transport mode of given origin and destination based on travel time [23], and commuting transport mode based on weak-labelling of visited cell towers [24]. Due to the challenging nature of the problem, there are very few studies reported on using CDR data in the route assignment step, which is the most detailed level of the transportation in all four steps. These studies include a simulation-based approach for route choice estimation by Tettamanti et al. [25], but with drawbacks in its feasibility in real-world scenarios where CDR data is spatially more coarse-grained and sparse than in its simulated situation. Another work is by Breyer et al. [26], which is an approach to reconstruct used routes based on CDRs; however, its shortcomings are the case that only one route can be estimated per visited cellular zone. The last is a work by Bwambele et al. [27] is an attempt to model route choice behavior, but for long-distance trips at an inter-regional level. While other previous studies have captured commuting patterns at a zonal scale such as clustered areas [28], cellular tower locations [29], and grid cells [30], this study attempts to extend the literature and fill in the gap by proposing and evaluating models for inferring commuting route choice at street level based solely on a CDR data.

6.2　Methodology

6.2.1　Dataset

A CDR is a collection of individual mobile phone communication logs gathered for billing purposes by a telecom operator. A communication log is kept each time a mobile phone user connects to his or her subscribed cellular network by making or receiving a call. Caller ID, Receiver's ID, caller's linked cell tower ID, Receiver's connected cell tower ID, timestamp, and call duration are all included in each report. The serving cell tower, which is the closest tower or base station to the user, is referred to as a linked cell tower.

　CDR data from 1.8 million mobile phone customers in Portugal over the period of one calendar year, from April 2006 to March 2007 was used. Individual phone numbers were anonymized by the operator before leaving their storage facilities to protect personal information, and were identifiable with a security ID (hash code) that complied with the EU GDPR. Data processed on a secure machine, in accordance with Article 89 of the GDPR, which permits the processing of personal data for research purposes. We chose a group of mobile users whose cellular network connections were at least five times a month for our study because we were interested in analysing the users' mobility. Communication activities were observed for each of the 12 months to ensure a fine-grained mobility observation. After filtering, we found 110,213 users who were the subjects of our study [31]. Our dataset contains a total of 6511 cell towers. Each cell tower has a unique identifier that corresponds to its geographic location (latitude and longitude).

6.2.2　Residence and Work Location Inference

The commuting excursion, whose origin and destination are residence and workplace, piqued our interest. As a result, the first piece of information we needed to study in order to explore the route choice was the location of each subject's home and workplace. During the sleeping hours (10:00 p.m.–7:00 a.m.), we employed the approach used in prior studies [22, 31, 32] to infer each individual subject's home location as the cell tower location that was used most frequently (highest connectivity) during the 12-month period. Similarly, the location of the workplace is deduced from the location of the cell tower that was most frequently used during weekday office hours (9 a.m.–5:00 p.m.).

6.2.3　Route Choices

After determining each subject's home and work locations, the Google Maps Directions API is used to generate a list of possible route options. The Google Maps Directions API gives us a set of plausible realistic route alternatives with our supplied set of origin and destination, which are the inferred home and workplace

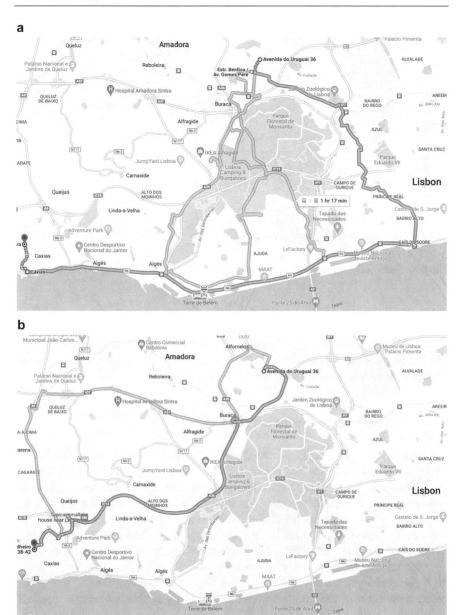

Fig. 6.1 Using the Google Maps Directions API, examples of various route choices for (**a**) driving and (**b**) public transportation

locations [33–36], because there are numerous route options between home and workplace. For each subject, we used the API to obtain possible travel options via vehicle and public transportation. Figure 6.1 shows an example of different route options provided in Google Maps. We received a collection of waypoints—that is, a

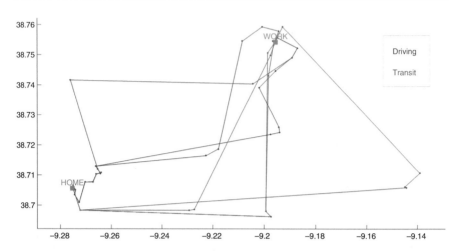

Fig. 6.2 Route options created using the Google Maps Directions API's received waypoints data

sequence of geo-coordinates along the route choice—in response to our HTTP request for route options using the Google Maps Directions API. Depending on the route's geometry, each route option may have a varied number of waypoints. There are more waypoints on curvier paths. Figure 6.2 depicts the waypoints obtained from the route selections presented in Fig. 6.1.

6.2.4 Route Choice Inference Methods

Our task was to determine the potential route options that were most likely taken by the individual based on his or her mobile phone usage pattern. Figure 6.3 shows an example of a subject, where three different routes between home and workplace are shown. Each yellow circle represents a used/connected cell tower location over the course of a year, with the size of the circle corresponding to the total number of connections—the larger the circle, the more frequently visited the place. Our task can be formulated as finding the route that is closest to the circles (visited locations).

Intuitively, our strategy is to evaluate the possibility of a route being chosen based on a calculated value measured between it and all visited cell towers, while also taking the number of visits into account. Let r k signify a collection of route choice k waypoints, i.e., $r_k = \{x_k(1), x_k(2), x_k(3), \ldots, x_k(M)\}$, where each waypoint i, i.e., $x_k(i) = \{Lat_k(i), Lng_k(i)\}$ contains a pair of geo-location coordinates; latitude and longitude. The likelihood score of route k (Wk) is the total ratio of the number of visits to the geographical distance between route k and each visited tower location.

$$W_k = \sum_{i=1}^{M} w_k(i), \tag{6.1}$$

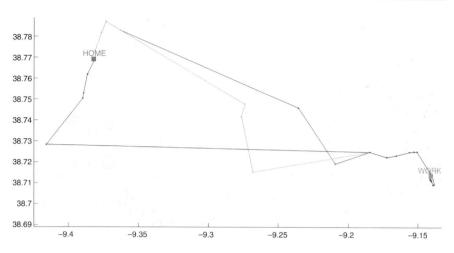

Fig. 6.3 shows an example of route options between home and work, as well as the locations of connected cell towers, with the circle size corresponding to the total number of connections

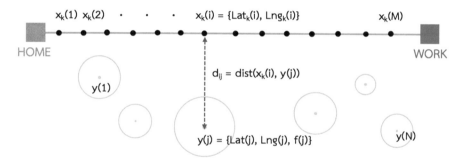

Fig. 6.4 Distance computation graphics between a route option and visited cell tower locations

where M is the total number of waypoints and $w_k(i) = \sum_{j=1}^{N} f(j)/d_{ij}$, where N is the total number of visited cell towers, $f(j)$ is the total number of visits to cell tower j, and $d_{ij} = dist(x_k(i), y(j))$ is the geographical distance (in kilometers) between a waypoint i and a cell tower j, as follows.

$$d_{ij} = 2R \cdot arcsin$$

$$\times \left(\sqrt{sin^2\left(\frac{Lat(j) - Lat_k(i)}{2}\right) + cos(Lat_k(i))cos(Lat(j))sin^2\left(\frac{Lng(j) - Lng_k(i)}{2}\right)} \right),$$

$$(6.2)$$

where $y(j) = \{Lat_k(i), Lng_k(i), f(j)\}$ are the geo-coordinates of cell tower j, and R is the Earth radius (6371 km). Figure 6.4 depicts the distance estimation method, with black dots representing waypoints between home, workplace and circles representing visited cell tower locations. Essentially, the route with the highest likelihood score (W_k) is chosen as the preferred option among all alternatives, i.e., $\underset{k}{\text{argmax}}\, W_k$.

6.2.4.1 Interpolation-Based Method

Depending on the nature of the route, the number of waypoints for each route option may vary. More waypoints are found on curvier routes. As a result, a route with more waypoints may have a higher likelihood score (W_k) since it sums across a larger number of terms (M), or waypoints. An interpolation is used to equalize the number of waypoints on each route in order to make a more fair comparison between all route choice candidates. Each route can be separated into the same number of segments (or edges) with an equal number of waypoints, as shown in Fig. 6.5.

To obtain a collection of interpolated waypoints, we must first identify the edges or road segments that make up the entire route. Each edge connecting two adjacent interpolated waypoints comprises information on the adjacent waypoints' locations, as well as its slope, length, and total distance from the origin (i.e., home). Let $x_k(i)$ and $x_k(i + 1)$ denote consecutive waypoints of edge $e_{(i)(i + 1)}$, which has a slope of m_i and a cumulative distance of c_i from the origin. A graphic depicting the physical meaning of these variables is shown in Fig. 6.6. A collection of route's initial

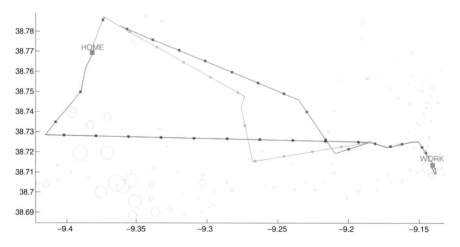

Fig. 6.5 shows an example of route options (given in Fig. 6.3) following interpolation, which equalizes the number of waypoints for all feasible routes

Fig. 6.6 Graphics depicting variables involved in determining a set of edges from a collection of route waypoints

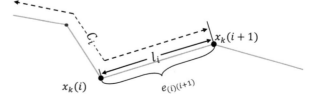

waypoints $r_k = \{x_k(1), x_k(2), x_k(3), \ldots, x_k(M)\}$ is given as an input, and the desired output is a set of route's edges. $E_k = \{e_{(1)(2)}, e_{(2)(3)}, \ldots, e_{(M-2)(M-1)}, e_{(M-1)(M)}\}$. Algorithm 6.1 describes the method of obtaining a set of route edges.

Algorithm 6.1 To obtain a set of route edges.
Input: Route waypoints, $r_k = \{x_k(1), x_k(2), x_k(3), \ldots, x_k(M)\}$
   Output: Set of route edges, E_k

1	**for** $i \leftarrow 1$ **to** $M - 1$ **do**
2	$m_i \leftarrow \dfrac{Lat_k(i+1) - Lat(i)}{Lng_k(i+1) - Lng(i)}$
3	$l_i \leftarrow dist(x_k(i), x_k(i+1))$
4	$C_i \leftarrow C_{i-1} + l_i$
5	$e_{(i)(i+1)} \leftarrow \{x_k(i), x_k(i+1), m_i, l_i, C_i\}$
6	**end for**
7	$E_k \leftarrow \{e_{(1)(2)}, e_{(2)(3)}, \ldots, e_{(M-2)(M-1)}, e_{(M-1)(M)}\}$
8	**return** E_k

Once a set of edges is obtained by using the Algorithm 6.1, we can then use this edge information to interpolate the route by adjusting the edge length and locations according to a required new number of edges. The process starts with a new interpolated edge length (l'), which can be simply calculated as a ratio of the whole route length and a required new number of edges (t) — i.e., the new number of road segments along the route. A new waypoint is assigned based on the original waypoint locations and the new interpolated edge length. Our method is described formally by the Algorithm 6.2.

Algorithm 6.2 To obtain interpolated route waypoints.
Input: Route edges (E_k) and number of interpolated edges (t)
   Output: Interpolated route waypoints, $x_k(i')$

1	**for** $i' \leftarrow 1$ **to** $t - 1$ **do**
2	$\Delta l \leftarrow i' \cdot l' - C_i$
3	$\theta \leftarrow tan^{-1}(m_i)$
4	**if** $Lng_k(i) < Lng_k(i+1)$
5	$Lng_k(i') = Lng_k(i) + \Delta l cos\theta$
6	$Lat_k(i') = Lat_k(i) + \Delta l sin\theta$
7	**else**
8	$Lng_k(i') = Lng_k(i) - \Delta l cos\theta$
9	$Lat_k(i') = Lat_k(i) - \Delta l sin\theta$
10	**end if**
11	$x_k(i') = \{Lat_k(i'), Lng_k(i')\}$
12	**end for**
13	**Return** $r'_k = \{x_k(1), x_k(2), \ldots, x_k(t-1)\}$

6.2.4.2 Shortest Distance-Based Method

Another approach to measuring distance from the route to the visited cell tower locations is to find the shortest distance between each visited cell tower location and each route segment or edge ($e_{(i)(i + 1)}$) instead of the distance to each waypoint ($x_k(i)$). Intuitively, this approach helps better reflect on a more realistic distance from the route to the visited cell tower.

With this approach, the distance d_{ij} in Eq. (6.2) can therefore be calculated as a distance from a visited cell tower location $y(j)$ that is perpendicular to an edge $e_{(i)}$ $_{(i + 1)}$ —i.e., shortest distance. If there is no perpendicular distance from $y(j)$ to $e_{(i)}$ $_{(i + 1)}$, then d_{ij} is a distance from $y(j)$ to either adjacent waypoint ($x_k(i)$ or $x_k(i + 1)$) whichever is the shortest. Figure 6.7 shows an illustrating graphic example of a case where there is a perpendicular distance from a visited cell tower location to the edge. On the other hand, Fig. 6.8 shows an example of another scenario where there is no

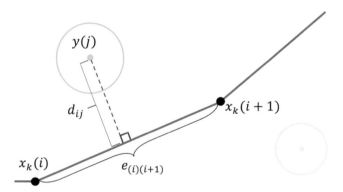

Fig. 6.7 Illustrating graphics of the shortest distance (d_{ij}) being a perpendicular distance from a visited cell tower location $y(j)$ to an edge $e_{(i)(i + 1)}$

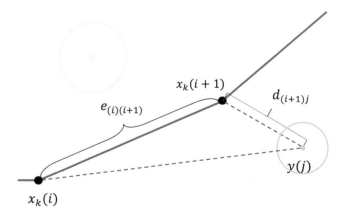

Fig. 6.8 Illustrating graphics of the shortest distance (d_{ij}) being a distance from a visited cell tower location $y(j)$ to the nearest waypoint ($x_k(i + 1)$ in this case) along the edge $e_{(i)(i + 1)}$

perpendicular distance along the edge, so the distance d_{ij} is calculated as a distance between the visited cell tower and the nearest waypoint of the considered edge.

Practically, as we operate in a discrete domain, locating a point along the edge that projects a perpendicular distance to a visited cell tower can be done approximately by interpolating the edge and finding the point with the shortest distance to the cell tower. The number of interpolated points along an edge can be set such that it gives a reasonable separation between the points for which we used one-meter spacing. Hence, number of interpolated points along an edge can be calculated as $\lceil dist(x_k(i), (x_k(i+1)) \times 10^3 \rceil$ for each edge. With a given edge and its calculated number of interpolated points (t'), the Algorithm 6.2 can be applied to obtain the interpolated points along the edge from which the shortest distance can then be calculated. Algorithm 6.3 describes our method of obtaining a shortest distance, where $E'_{(i)(i+1)} = \{x'(1), x'(2), \ldots, x'(t')\}$ is a set of interpolated points along the edge i.

Algorithm 6.3 To find a shortest distance.
Input: Edge $e_{(i)(i+1)}$ and cell tower $y(j)$
    Output: Shortest distance, d_{ij}
    1     $t' \leftarrow \lceil dist(x_k(i), (x_k(i+1)) \times 10^3 \rceil$
    2     $E'_{(i)(i+1)} = Algorithm2(e_{(i)(i+1)}, t')$
    3   **for** $i \leftarrow 1$ **to** $t' - 1$ **do**
    4         $c_{ij} \leftarrow dist(x'(i), y(j))$
    5 **end fors**
    6 **Return** $\underset{i}{\text{argmin}} c_{ij}$

6.2.4.3 Voronoi Cell-Based Method

Voronoi diagram is a typical approach for partitioning space into sub-regions based on a collection of predefined points in the space. It is frequently employed in the disciplines of spatial analysis and urban planning, such as service area delimitation [37] and map generalization [38]. As it is for space partitioning based on the distance to a set of seed points, the Voronoi diagram can be directly applied in our case to partition the entire area into sub-areas (or Voronoi cells) based on the cell tower locations — i.e., generating a coverage zone of each cell tower – which can then be used as a spatial reference in measuring a distance from the route d_{ij}. Figure 6.9 displays the created Voronoi cells that establish service coverage zones across the country according to all 6511 cell tower locations.

With these generated Voronoi cells, a distance from a route to each visited cell tower can be calculated from the points on its Voronoi boundaries that the route passes through. Figure 6.10 demonstrates an example where the points on the visited cell tower's Voronoi boundaries are marked with black solid circles. As shown in Fig. 6.11, the distance between the route and each visited cell tower (or d_{ij} in Eq. 6.2) can be calculated in one of two ways: (1) the sum of distances from all passed points to the cell tower location i.e., $d_{ij} + d_{(i+1)j}$; or (2) the minimum distance among all

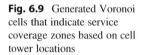

Fig. 6.9 Generated Voronoi cells that indicate service coverage zones based on cell tower locations

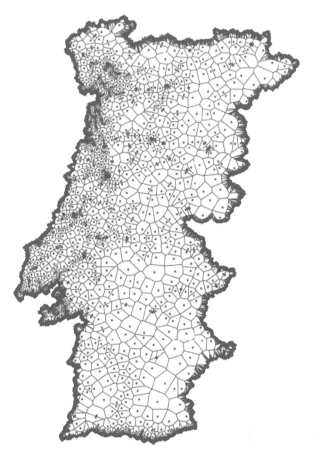

crossed points to the cell tower location i.e., $d_{(i + 1)j}$ in this example). Note that in this study, Voronoi cells were constructed using the Matlab function *voronoi*, and the intersection locations between route and Voronoi cell boundaries were found using the Matlab function *polyxpoly*.

6.2.4.4 Visited Voronoi Cell-Based Method

With the Voronoi-cells-based method, some segments of the route may not pass across visited Voronoi cells and hence are not taken into account when calculating distance. This can be solved by introducing a new approach in which the main concept is to start with an individual subject's Voronoi-cell map, which is constructed individually by solely taking into account the subject's visited cell tower locations. As a result, each of the route options will pass through visited Voronoi cells along the way, allowing the distance to be calculated over the entire route. Figure 6.12 displays an example of a sample subject's visited Voronoi-cell map (with a zoom-in of the Lisbon area), which generates the Voronoi cells based solely on the subject's visited cell tower locations. Figure 6.13 illustrates a sample

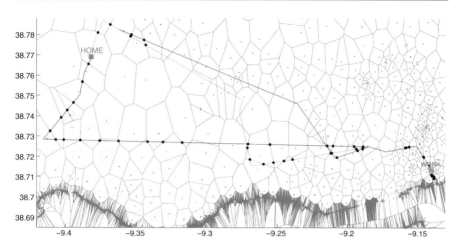

Fig. 6.10 shows spots on the visited cell tower's Voronoi borders that each route passes over (marked with black solid circles)

Fig. 6.11 Graphics demonstrating how the Voronoi cell-based method is used to calculate distance from the route and visited cell tower

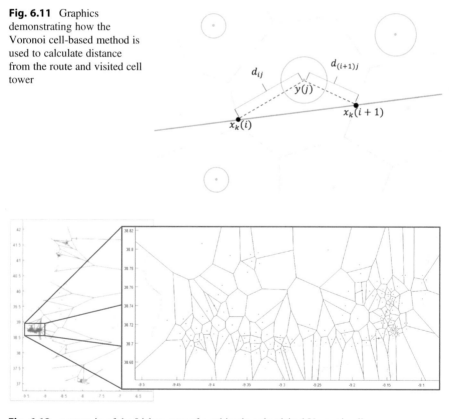

Fig. 6.12 a zoom-in of the Lisbon area of a subject's only visited Voronoi-cell map

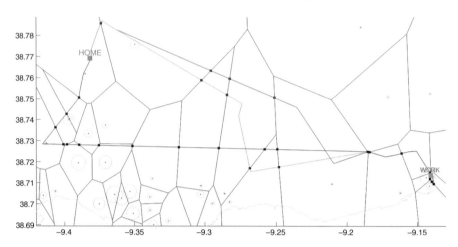

Fig. 6.13 Using the visited Voronoi-based method, each route passes through points (marked with black solid circles) on the visited cell tower's Voronoi boundaries

subject's (previously shown in Fig. 6.10) route choices that travel via the visited Voronoi cells using this method. The distance between each route and visited cell towers can then be calculated using the Voronoi-cell based method as previously mentioned (Sect. 6.2.4.3).

6.2.4.5 Noise Filtering

Although commuting trips account for the majority of individual journeys, there are also trips to and from locations other than home and work. These non-commuting journeys are also recorded in the form of connectivity logs and can be seen in the CDR data. The cellular network connectivity associated with these non-commuting journeys, however, can be regarded as noise because our focus is on commuting trips.

Here we introduce two approaches to reduce the noise in our commuting route choice inference. The first approach is using DBSCAN [39], which is a density-based spatial clustering method that groups data points with many adjacent neighbors and filters out outliers, which are data points that are isolated in low-density regions. The radius of a neighborhood and the minimum number of points required to construct a dense zone (*minPts*) are two required parameters for the DBSCAN, which we set to be equal to the commuting distance (i.e., direct distance between home and workplace) and *minPts* = 10, respectively. The parameters' values were chosen and justified based on our observations of the outcomes, which led us to conclude that our chosen values were appropriate. Different practical DBSCAN parameter selections could, of course, be interesting to investigate in a future study. Wong and Huang [40] conducted a sensitivity analysis of spatiotemporal trajectory data clustering and found that choosing the best values for these two parameters is still an open research subject. The two parameters appear to work against one another to some extent—raising *minPts*

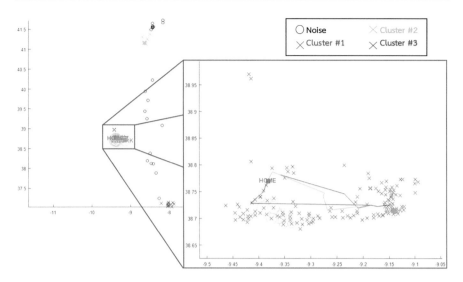

Fig. 6.14 Clustering result of DBSCAN from which it's Cluster #1 is further considered for the route choice inference while the rest of the visited cell towers are considered as a noise and discarded

disbands bigger clusters into smaller ones, but value defines the geographical scale of cluster identification, so increasing value increases extensive clusters. Some research [41, 42] have recommended appropriate values for these two factors, but these suggestions are not universal because they are data dependent. The starting point in our scenario is the subject's home cell tower location, hence the first cluster is a collection of visited cell tower locations throughout the commuting route. As a result, just this first cluster is taken into account in our route choice inference, while other clusters and noise are disregarded as a whole. Figure 6.14 displays a DBSCAN clustering result for a sample subject, with Cluster #1 being used in our route inference approach and the rest of the visited cell towers being disregarded as noise.

Our second approach to filtering out visited cell towers that are unlikely related to commuting trip, i.e., cell towers that are further away from the route choices, or considered as noise involves the use of commuting distance, i.e., direct home-workplace distance, which is used as 'commuting radius' to draw a noise filtering scope, as an analogy to bandpass filter. This noise filtering scope is centered at the midpoint on a straight line drawn between the home and workplace. All visited cell towers that are located within the scope of commuting radius are then further considered for our route inference while the rest is considered as noise and discarded. Figure 6.15 shows a graphic demonstrating this commuting-radius based noise filtering approach. For an actual example, Fig. 6.16 shows a result of the commuting-radius based noise filtering (of the subject previously shown in Fig. 6.14) where cell towers located within the enclosed commuting-radius scope are further taken otherwise discarded as noise.

Fig. 6.15 Illustrating
graphics showing how the
commuting-radius based
noise filtering works. Cell
towers located within the
commuting-radius (dc) scope
are taken further for the route
choice inference while those
located outside the scope are
considered as noise and
discarded

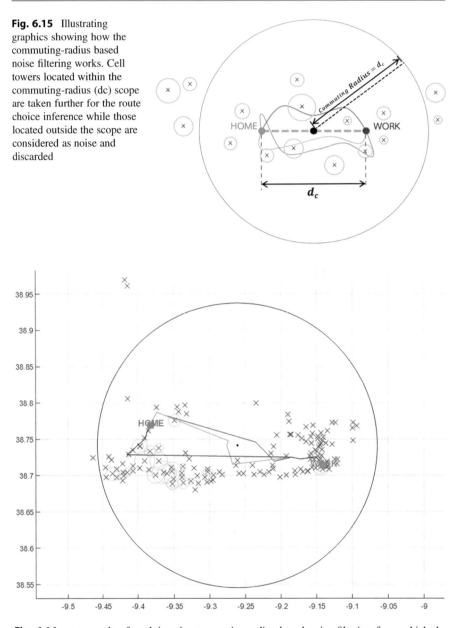

Fig. 6.16 an example of applying the commuting-radius based noise filtering from which the visited cell towers located within the commuting-radius scope are taken further for the route inference but those located outside the scope are considered as noise and then discarded

6.3   Results

Here, we present the results of our route inference methods including the interpolation-based, shortest distance-based, Voronoi cell-based, and visited Voronoi cell-based methods implemented with and without noise filtering (DBSCAN and commuting radius-based) (DBSCAN and commuting radius-based). Accuracy rate was calculated by comparing the inferred route against a ground truth.

Since obtaining the actual route choice information from the subjects was not possible, a ground truth based on a visual inspection and hand labelling of data was generated. This was believed to be most reasonably feasible in the case here. For consistency, one person was designated for the task and asked to hand label the route that was mostly believed to be the taken commuting route choice. The designated hand-labelling person viewed a subject's CDR connectivity along with route choices (similar to the one shown in Fig. 6.3) and was asked to identify the most probable route taken. The hand-labelling person was asked to only hand label the most probable route with high confidence, so the person did not hand label every examined subject but only those whose most probable route choices were clearly obvious to her. This exhaustive hand-labelling task yielded a 90-subject ground truth for our experiment.

6.3.1   Interpolation-Based Methods

For the interpolation-based method, an accuracy rate was calculated for each of the varying number of interpolated edges ranging from 2 to 100 edges to observe the impact of the level of interpolation. Furthermore, the used (visited) cell towers were ranked from the most to the least used towers, for which an accuracy rate was calculated from the top 1% to 100% (all) used cell towers.

Overall result is shown in Fig. 6.17a for a total of $99 \times 100 = 9900$ experimental setups for which accuracy rates were calculated. The overall average accuracy rate is 82.05%. The accuracy reaches its maximum of 90% for 62 times which all happen when the number of interpolated edges is set to 12 edges and the top cell tower percentage varies from 34% to 95%. Figure 6.17b shows the average accuracy rates of each of the varying number of interpolated edges along with corresponding standard deviation bars, which confirms that with 12 interpolated edges, the average accuracy rate is at the highest of 87.72%, averaged across all top cell tower percentage variations. Interestingly, when considering the average accuracy rates across all top cell tower percentages as shown in Fig. 6.17c, the average accuracy gradually rises and becomes stable as the number of top cell towers considered for accuracy rate calculation increases. The average accuracy rate rises to 83.04% at 26% top cell towers and does not change much as it continues to slightly climb up to 83.79% when it reaches all 100% top cell towers. This suggests that only some percentage of top visited cell towers can be sufficient for the route choice inference, as it does not significantly improve the accuracy by taking more data presumably

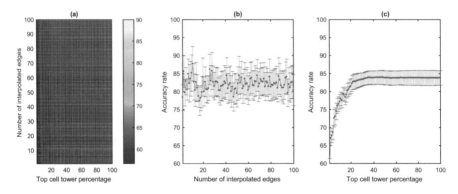

Fig. 6.17 Results based on the interpolation-based method: (**a**) overall accuracy rates; (**b**) average and standard deviation of the accuracy rates of varying numbers of interpolated edges; (**c**) average and standard deviation of the accuracy rates of varying numbers of top cell tower percentage

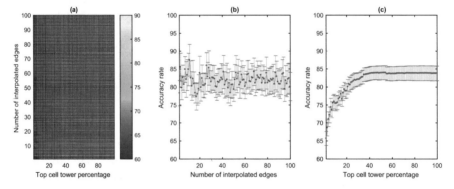

Fig. 6.18 Results based on the interpolation-based method implemented with DBSCAN-based noise filtering: (**a**) accuracy rates; (**b**) average and standard deviation of the accuracy rates of varying numbers of interpolated edges; (**c**) average and standard deviation of the accuracy rates of varying numbers of top cell tower percentage

less relevant. This result opens up an interesting research question of how much of the individual CDR data that is said to be significantly sufficient and relevant to the route choice inference, which is worth future investigation.

For the interpolation-based method implemented with the DBSCAN-based noise filtering, the examined number of interpolated edges varies from 2 to 100 edges, while the percentage of top cell towers ranges from 2% to 100% in this experiment (as there were not top cell towers in some cases made up by 1% after some towers being filtered out as noise). The overall result from a total of $99 \times 99 = 9801$ experimental setups is shown in Fig. 6.1a. The overall average accuracy is 81.85%, while the highest accuracy rate is 90% when the number of interpolated edges is 12 and the percentage of top cell towers is from 37% to 95%. Along the same line, Fig. 6.18b shows that the highest average accuracy rate of 87.49% is reached when 12 interpolated edges were used. Our examination of top cell tower percentages in

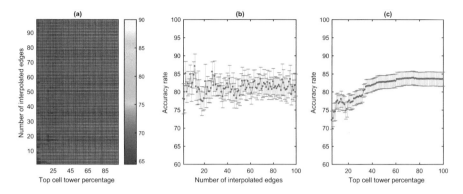

Fig. 6.19 Overall results based on the interpolation-based method with commuting radius-based noise filtering; (**a**) accuracy rates, (**b**) average accuracy rates with varying numbers of interpolated edges, and (**c**) average rates with varying percentages of top cell towers

Fig. 6.18c shows that with DBSCAN, the average accuracy rate increases slightly slower than the interpolation-based method without DBSCAN to reach its stable level. In this experiment, it reaches a stable level (83.41% accuracy) at 35% of top cell towers, which is slower than that of the normal method whose stable accuracy rate is reached at 26% of top cell towers. The average accuracy once reaches its stable level at 83.41%; it continues to slowly rise to 83.79% when the entire cell towers are considered.

Lastly, with the commuting radius-based noise filtering, the interpolation-based method performs slightly worse than the previous two methods. The variation of examined number interpolated edges is the same as in previous experiments which is 2–100 edges, but the percentage of top cell towers in this experiment varies from 6% to 100%. The overall result obtained from a total of 99 × 96 = 9504 experimental setups is shown in Fig. 6.19a where the overall average accuracy rate is 81.35%. The highest accuracy is 90%, which happens when the number of interpolated edges is 12 and the percentage of top cell towers is from 59% to 93%. From the perspective of the number of interpolated edges, the average accuracy reaches its maximum at 84.70% with 12 edges, as shown in Fig. 6.19b. With the varying percentage of top cell towers, the accuracy reaches its stable rate at 83.03% when the top 58% cell towers were considered, and it moves up and down slightly and eventually stands at 83.59% when the entire visited cell towers were taken into consideration.

6.3.2 Shortest Distance-Based Methods

With the shortest distance-based method, an accuracy rate was calculated for each varying percentage of top cell towers implemented with and without noise filtering. All results from three different models are shown in Fig. 6.20, including the shortest distance-based method without noise filtering (*Shortest distance*), the shortest distance-based method with DBSCAN (*Shortest distance + DBSCAN*), and the

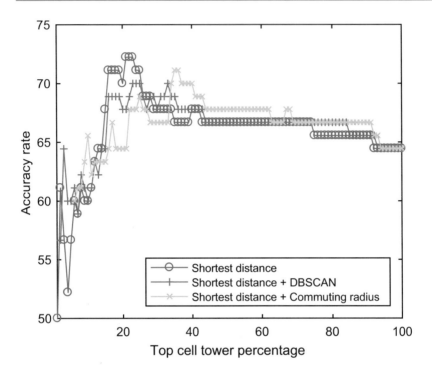

Fig. 6.20 Results of the shortest distance-based methods with and without noise filtering

shortest distance-based method with commuting radius-based noise filtering (Shortest distance + Commuting radius). The top cell tower percentage varies from 1–100% for the shortest distance method, from 2–100% for the shortest distance + DBSCAN method, and from 6–100% for the *Shortest distance + Commuting radius* method.

With the *Shortest distance* method, its accuracy rate has an uprising trend and reaches the maximum value of 72.22% when the top cell lower percentage is 21%, 22%, and 23%. It then shows a continuous dropping trend to reach 64.44% accuracy rate when the entire (100%) visited cell towers were taken into account. The result of the *Shortest distance + DBSCAN* method shows a similar up-and-down trend with the shortest distance method. It rises to reach its maximum at 70% when the top cell tower percentage is 23%, 24%, and 33%, then gradually drops to eventually reach 64.44% when the entire cell towers are considered. Lastly, the resulting accuracy rates of the *Shortest distance + Commuting radius* method also appear to show a similar trend where it rises to its maximum value of 71.11% when 35% or 36% top cell towers were considered. It then slowly decreases to eventually reach 64.44%. Overall, the shortest distance-based method without noise filtering appears to have better performance than implementing it with noise filtering, as it shows to be the fastest to reach its maximum accuracy rate, and it also poses the highest accuracy value among all three examined methods.

6.3.3   Voronoi Cell-Based Methods

The Voronoi cell-based method was implemented with and without noise filtering. The results are shown in Fig. 6.21. Percentage of top cell towers considered for the route inference varies and starts from 15%, 22%, and 35% for the method without noise filtering (*Voronoi*), DBSCAN (*Voronoi + DBSCAN*), and with commuting radius-based noise filtering (*Voronoi + Commuting radius*), respectively. The top cell tower percentage starts from a different value in each of the three methods due to the obtainable amount of top cell towers with a specified percentage value. Two separated set of experiments were implemented, one with the summed distance approach and the other with the shortest distance approach, as described in Sect. 6.2.4.3—i.e., distance measured from passed points to a visited cell tower location (d_{ij}, $d_{(i + 1)j}$ to $y(j)$). For the summed distance approach as shown in Fig. 6.21a, the method shows an uprising trend to reach its maximum accuracy of 78.89% with top 26% cell towers and remains at this accuracy until the top 54% cell towers considered, then it drops slightly to 77.78% and remains there from top 55–82% cell towers and then takes another step drop to 76.67% and lasts for the rest of the cell tower percentages. The *Voronoi + DBSCAN* method exhibits a similar result with the *Voronoi* method, as its accuracy rate climbs up in the same fashion but reaches the same maximum rate (78.89%) later at top 29% cell towers and remains at the level before it drops to 77.78% at the same 55% top cell tower level but remains there until 85% top cells before taking the last step drop to the same accuracy rate of 76.67% for the rest of the way. The three-step accuracy rates are also observed for the *Voronoi + Commuting radius* method, as it rises to the maximum (78.89%) and remains there from 42% to 75% top cells, and then later down steps are 76–94% and 95–100%. These stepwise accuracy rates are most possibly due to the issue previously discussed in Sect. 6.2.4.3 that with the Voronoi-cells based method, there may potentially be some portions of the route that do not pass through visited Voronoi

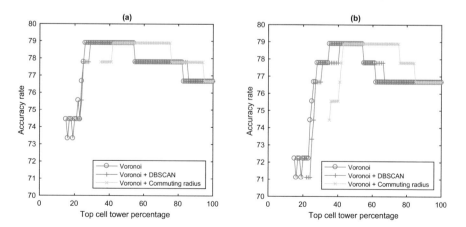

Fig. 6.21 Results of the Voronoi cell-based methods with and without noise filtering: (**a**) summed distance approach; (**b**) shortest distance approach

cells and hence they are not considered in distance calculation. A chunk of top cell towers are likely required to make an impact on accuracy rate, thereby a stepwise accuracy result is observed.

For the Voronoi-based shortest distance approach, the result is shown in Fig. 6.21b where the top cell tower percentage starts at 15%, 22%, and 35% for the Voronoi cell-based method without noise filtering (*Voronoi cells*), with DBSCAN (*Voronoi + DBSCAN*), and with commuting radius-based noise filtering (*Voronoi + Commuting radius*), respectively. Stepwise accuracy results are also observed here. The calculated accuracy rates of all three models exhibit a similar trend, as they all rise to reach the same maximum rate of 78.89% and remain there for some top cell tower percentages, then drops to 77.78% and later 76.67%. The *Voronoi* method reaches the maximum with the least percentage of top cell towers of 35%, followed by the *Voronoi + DBSCAN* of 41% and then *Voronoi + commuting radius* of 43%. The *Voronoi + Commuting radius*, however, is able to stay at the maximum rate for the largest portion of the top cell percentages, i.e., 43–75%.

6.3.4   Visited Voronoi Cell-Based Methods

Likewise, the visited Voronoi cell-based method (*V-Voronoi*) was implemented without and with noise filtering (*V-Voronoi + DBSCAN* and *V-Voronoi + Commuting radius*), as well as altered with the summed distance and shortest distance approaches. The top cell percentage starts at 11%, 12%, and 39% for *V-Voronoi*, *V-Voronoi + DBSCAN*, and *V-Voronoi + Commuting radius*, respectively. The results are shown in Fig. 6.22.

With the summed distance approach, the maximum accuracy among three methods of 75.56% was achieved by the *V-Voronoi + Commuting radius* at 42% top cell tower level, followed by the *V-Voronoi* at 73.33% when top cell percentage

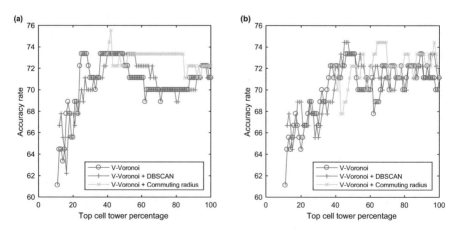

Fig. 6.22 Results of the visited Voronoi cell-based methods with and without noise filtering; (**a**) summed distance approach and (**b**) shortest distance approach

is in the ranges 25–28% and 36–49%, and the *V-Voronoi + DBSCAN* at also 73.33% for a span of 40–56% top cell towers. With the shortest distance approach, the maximum accuracy among all three methods was 74.44%, which was achieved by *V-Voronoi + DBSCAN* for 45–47% top cell towers and *V-Voronoi + Commuting radius* for 64–69% top cell towers.

Interestingly, the performances of the visited Voronoi cell-based methods are lower than that of the Voronoi cell-based methods. This may be due to the consideration of only the visited Voronoi cells that draw up larger coverage zones (or cells), which may negatively affect the distance calculation.

6.3.5 Result Summary

The results from all sets of experiment of our developed and examined methods for route choice inference including the interpolation-based, shortest distance-based, Voronoi cell-based, and visited Voronoi cell-based methods, implemented with and without DBCAN or commuting radius-based noise filtering are summarized in Table 6.1.

Interestingly, the interpolation-based method has the best result in both points of view of the maximum and average accuracy rates. From the average accuracy rate's perspective, the top five rankings are Interpolation (82.05%), Interpolation + DBSCAN (81.85%), Interpolation + Commuting radius (81.35%), V-Voronoi cells (summed dist.) + Commuting radius (77.88%), and then Voronoi cells (shortest dist.) + Commuting radius (77.83%). The bottom five rankings include shortest distance (65.89%), shortest distance + DBSCAN (66.20%), Shortest distance + Commuting radius (66.61%), V-Voronoi cells (shortest dist.) (70.12%), and V-Voronoi cells (shortest dist.) + DBSCAN (70.46%). If grouped by the main method, the interpolation-based group has the highest average accuracy rate (81.75%), followed by the Voronoi cells (summed dist.) group (77.54%), and Voronoi cells (shortest dist.) group (77.25%).

From the point of view of a receiver operating characteristic curve, or ROC curve, which is a performance measurement for classification problem at various thresholds settings, the performance across all route inference models with varying top cell percentages is in line with the results observed in Table 6.1 as all interpolation-based models are among the top performance on the ROC curve, as shown in Fig. 6.23 Model performance is quantified in forms of true positive rate (TPR) or sensitivity versus false positive rate (FPR) or chance of false alarm.

Since our ultimate purpose of this inquiry is to offer a new and more efficient alternative than the usual surveys to obtain route choice information at a wide scale, we utilized our best method (i.e., interpolation-based with 12 edges) to infer commuting route choices in Portugal. Our ground truth of 90 individuals' route choices is given in Fig. 6.24a, and the inferred route choices of 110,213 subjects on the Portugal road network depicted in Fig. 6.24b. With our approach, commuting route choice information can be gathered at any time, which is intuitive and more up-to-date compared to the travel survey that is collected once every 5–10 years. As

Table 6.1 Result summary of all proposed commuting route choice inference methods

Method	Accuracy			Top cell tower percentage	Number of interpolated edges
	Max	Min	Avg.		
Interpolation	**90.00**	56.67	82.05	1–100	2–100
Interpolation + DBSCAN	90.00	60.00	81.85	2–100	2–100
Interpolation + commuting radius	90.00	64.44	81.35	6–100	2–100
Shortest distance	72.22	50.00	65.89	1–100	–
Shortest distance + DBSCAN	70.00	56.67	66.20	2–100	–
Shortest distance + commuting radius	71.11	60.00	66.61	6–100	–
Voronoi cells (summed dist.)	78.89	73.33	77.54	15–100	–
Voronoi cells (summed dist.) + DBSCAN	78.89	74.44	77.82	22–100	–
Voronoi cells (summed dist.) + commuting radius	78.89	76.67	77.25	35–100	–
Voronoi cells (shortest dist.)	78.89	71.11	76.83	15–100	–
Voronoi cells (shortest dist.) + DBSCAN	78.89	71.11	77.09	22–100	–
Voronoi cells (shortest dist.) + commuting radius	78.89	74.44	77.83	35–100	–
V-Voronoi cells (summed dist.)	73.33	61.11	70.58	11–100	–
V-Voronoi cells (summed dist.) + DBSCAN	73.33	62.22	70.67	12–100	–
V-Voronoi cells (summed dist.) + commuting radius	75.56	71.11	77.88	39–100	–
V-Voronoi cells (shortest dist.)	73.33	61.11	70.12	11–100	–
V-Voronoi cells (shortest dist.) + DBSCAN	74.44	64.44	70.46	12–100	–
V-Voronoi cells (shortest dist.) + commuting radius	74.44	67.78	72.19	39–100	–

a result, transportation planning, planning, and design can be better educated, planned, and designed to suit current traffic demand and travel behavior.

6.4   Conclusion

Due to its widespread use, the mobile phone has evolved into a personal sensor that captures digital traces of individuals through the use of services such as voice conversations, text messaging, and the internet. These communication logs are gathered for billing purposes, however on a broad scale, these individual traces can be regarded as a significant behavioural data source for gaining insights into human behaviour. The location histories of mobile phone users are used in this study

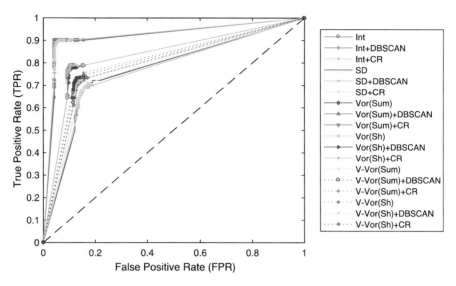

Fig. 6.23   ROC curve of each model's performance using top cell percentage for threshold settings

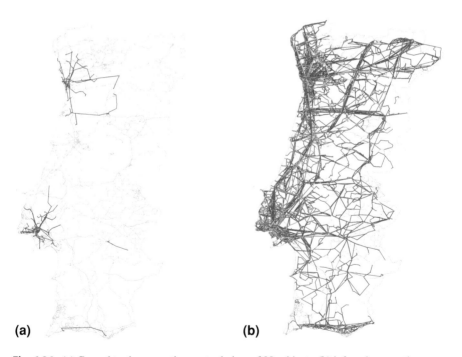

Fig. 6.24   (**a**) Ground-truth commuting route choices of 90 subjects, (**b**) inferred commuting route choices of 110,213 subjects on the Portugal road network

to obtain information on people's commuting route choices. This is a more cost-effective and time-consuming alternative to standard travel surveys such as roadside interviews and questionnaires. A total of 18 distinct route choice inference models have been developed and examined. Based on the geographical traces of individual mobile phones' connectivity, i.e., connected (or visited) cellular towers, our route choice inference has been defined as a problem of choosing the most likely path taken among different route choice options. The shortest distance between a route choice with original waypoints and visited cellular towers, the Voronoi cells that assign a route choice into multiple coverage zones, and the consideration of only visited Voronoi cells that assign a route choice into multiple coverage zones are all based on interpolation of route waypoints for calculating distance between a probable route choice and connected cell towers. Two approaches to estimating the distance have been implemented for both Voronoi-based models: one uses the accumulated distance of passed points, while the other uses simply the shortest distance of all passed points on each evaluated Voronoi boundary. Each of these models has been implemented with and without noise filtering, including using the DBSCAN algorithm and our own method of drawing a noise filtering scope using commuting distance as a bandpass filter analogy. A mobile network data (i.e., CDR) from Portugal was used in the developing and examining of the models. As a result, the interpolation-based models perform best in terms of accuracy rate, followed by the consideration of only visited Voronoi cell-based model with summed distance approach and commuting radius-based noise filtering, and finally the Voronoi cell-based shortest distance approach and commuting radius-based noise filtering.

Reflecting on our findings, it appears that the noise may not be the 'noise' we previously thought. When all of the data is considered instead of just the noise filtered data, the results are usually better. As we varied the percentage of top (used/ visited) cellular towers in our accuracy calculation, relevance is probably the key here, as evidenced by the top cellular tower percentage experiments, where the model performs better with some portion of top cell towers than when the entire data is considered.

The commute route choices of people in Portugal are inferred and illustrated on the road network as a case study using a small ground truth set. The developed and examined models can have an immediate implication in gathering route choice information on a large scale that facilitates the four-step models in transportation forecasting. Since they provide a perspective on both issue genesis and solution, the models and methodologies can also serve as a baseline for future development and investigation into the route choice inference problem based simply on CDR data.

Certainly, our research does, however, have some limitations. The possibility of many route options is the first of these. In this study, we assumed that a person only uses one mode of transportation for commuting, which is not always the case, but probably is for the majority of individuals. Another potential drawback is from the exsausive nature of our hand labelling task, which results in an allegedly small set of ground truth. A larger ground truth size could potentially boost the model evaluation's significance. A further limitation is that the route choice candidates acquired using the Google Maps Directions API may not be the exact route options available. Nonetheless, we believe that the gathered route options are still very comparable to the actual ones, as the suggested route options are mostly based on the shortest travel time, which is likely the method taken by most commuters. As our study was limited to mobile phone users in Portugal, a final limitation is the extent to which our findings are transferable outside of the case study nation. Despite the fact that this is a case study to demonstrate our model development and analysis, the findings are still valid and can be applicable to countries with broadly similar social, cultural, and economic profiles to Portugal, which is a member of the Schengen countries and a developed country with significant similarities to several European and Asian countries.

In terms of future research directions, this study raises a lot of intriguing relevant research topics that could be investigated further in the future, such as how to select an appropriate percentage of top visited cell towers for route choice inference. Data streams, especially in the big data era, contain both important and useless elements. The goal is to figure out how to extract only the ones that are important. Another future research direction is to use the CDR data's temporal context in route choice inference, which is not covered in this chapter. The model's performance could be improved if temporal sequences are taken into account. Finally, the motivation behind route selections is an intriguing aspect to investigate, as it may be tied to the road network's connection and architecture.

References

1. Vickrey W. Congestion theory and transport investment. Am Econ Rev. 1969;59:251–60.
2. Mcnally MG. The four step model. Handb Transp Model. 2007;
3. Stopher PR, Greaves SP. Household travel surveys: where are we going? Transp Res Part A Policy Pract. 2007;21(5):367–81.
4. Shen L, Stopher PR. Review of GPS travel survey and GPS data-processing methods. Transp Rev. 2014;
5. Van Alsenoy B. General data protection regulation. In: Data protection law in the EU: roles, responsibilities and liability. 1st ed. KU Leuven Centre for IT & IP Law Series; 2019.
6. Cuttone A, Lehmann S, González MC. Understanding predictability and exploration in human mobility. EPJ Data Sci. 2018;
7. Rupi F, Poliziani C, Schweizer J. Data-driven bicycle network analysis based on traditional counting methods and GPS traces from smartphone. ISPRS Int J Geo-Information. 2019;8(8):322.
8. Phithakkitnukoon S, Horanont T, Witayangkurn A, Siri R, Sekimoto Y, Shibasaki R. Understanding tourist behavior using large-scale mobile sensing approach: a case study of mobile phone users in Japan. Pervasive Mob Comput. 2015;18
9. Caceres N, Wideberg JP, Benitez FG. Review of traffic data estimations extracted from cellular networks. IET Intell Transp Syst. 2008;2(3):179–92.
10. Blondel VD, Decuyper A, Krings G. A survey of results on mobile phone datasets analysis. EPJ Data Science. 2015;
11. de Montjoye Y-A, Hidalgo CA, Verleysen M, Blondel VD. Unique in the crowd: the privacy bounds of human mobility. Sci Rep. 2013;3(1):1376.
12. Shi F, Zhu L. Analysis of trip generation rates in residential commuting based on mobile phone signaling data. J Transp Land Use. 2019;12(1):201–20.

13. Bwambale A, Choudhury CF, Hess S. Modelling trip generation using mobile phone data: a latent demographics approach. J Transp Geogr. 2019;76:276–86.
14. Di Donna SA, Cantelmo G, Viti F. A Markov chain dynamic model for trip generation and distribution based on CDR. In: 2015 International Conference On Models and Technologies for Intelligent Transportation Systems; 2015. p. 243–50.
15. Bonnel P, Hombourger E, Olteanu-Raimond AM, Smoreda Z. Passive mobile phone dataset to construct origin-destination matrix: potentials and limitations. Transp Res Procedia. 2015;11: 381–98.
16. Demissie MG, Phithakkitnukoon S, Sukhvibul T, Antunes F, Gomes R, Bento C. Inferring passenger travel demand to improve urban mobility in developing countries using cell phone data: a case study of Senegal. IEEE Trans Intell Transp Syst. 2016;17(9):2466–78.
17. Wu H, Liu L, Yu Y, Peng Z, Jiao H, Niu Q. An agent-based model simulation of human mobility based on Mobile phone data: how commuting relates to congestion. ISPRS Int J Geo-Information. 2019;8(7):313.
18. Mamei M, Bicocchi N, Lippi M, Mariani S, Zambonelli F. Evaluating origin–destination matrices obtained from CDR data. Sensors. 2019;19(20):4470.
19. Hankaew S, Phithakkitnukoon S, Demissie MG, Kattan L, Smoreda Z, Ratti C. Inferring and modeling migration flows using mobile phone network data. IEEE Access. 2019;7(1): 164746–58.
20. Demissie MG, Phithakkitnukoon S, Kattan L. Understanding human mobility patterns in a developing country using Mobile phone data. Data Sci J. 2019;18(1):1–13.
21. Demissie MG, Phithakkitnukoon S, Kattan L. Trip distribution modeling using Mobile phone data: emphasis on intra-zonal trips. IEEE Trans Intell Transp Syst. 2019;20(7):2605–17.
22. Phithakkitnukoon S, Sukhvibul T, Demissie M, Smoreda Z, Natwichai J, Bento C. Inferring social influence in transport mode choice using mobile phone data. EPJ Data Sci. 2017;
23. Wang H, Calabrese F, Di Lorenzo G, Ratti C. Transportation mode inference from anonymized and aggregated mobile phone call detail records. In: 13th International IEEE Conference on Intelligent Transportation Systems, Proceedings (ITSC); 2010. p. 19–22.
24. Graells-Garrido E, Caro D, Parra D. Inferring modes of transportation using mobile phone data. EPJ Data Sci. 2018;7:49.
25. Tettamanti T, Demeter H, Varga I. Route choice estimation based on cellular signaling data. Acta Polytech Hungarica. 2012;9(4):207–20.
26. Breyer N, Gundlegård D, Rydergren C. Cellpath routing and route traffic flow estimation based on cellular network data. J Urban Technol. 2018;25(2):85–104.
27. Bwambale A, Choudhury C, Hess S. Modelling long-distance route choice using mobile phone call detail record data: a case study of Senegal. Transp A Transp Sci. 2019;15(2):1543–68.
28. Yang X, Fang Z, Yin L, Li J, Zhou Y, Lu S. Understanding the spatial structure of urban commuting using mobile phone location data: a case study of Shenzhen, China. Sustain. 2018;10(5):1435.
29. Jundee T, Kunyadoi C, Apavatjrut A, Phithakkitnukoon S, Smoreda Z. Inferring commuting flows using CDR data. In: Proceedings of the 2018 ACM International Joint Conference and 2018 International Symposium on Pervasive And Ubiquitous Computing and Wearable Computers; 2018. p. 1041–50.
30. Zagatti GA, et al. A trip to work: estimation of origin and destination of commuting patterns in the main metropolitan regions of Haiti using CDR. Dev Eng. 2018;3:133–65.
31. Phithakkitnukoon S, Smoreda Z, Olivier P. Socio-geography of human mobility: a study using longitudinal mobile phone data. PLoS One. 2012;7(6):e39253.
32. Horanont T, Phiboonbanakit T, Phithakkitnukoon S. Resembling population density distribution with massive mobile phone data. Data Sci J. 2018;17:1–9.
33. Chia WC, Yeong LS, Jia F, Lee X, Inn S. Trip planning route optimization with operating hour and duration of stay constraints. In: 2016 11th Int. Conf. Comput. Sci. Educ., no. Iccse; 2016. p. 395–400,

34. Chou YT, Hsia SY, Lan CH. A hybrid approach on multi-objective route planning and assignment optimization for urban lorry transportation. In: Proc. 2017 IEEE Int. Conf. Appl. Syst. Innov. Appl. Syst. Innov. Mod. Technol. ICASI 2017; 2017. p. 1006–9,

35. Nguyen H, Zhao H, Jamonnak S, Kilgallin J, Cheng E. RooWay: a web-based application for UA campus directions. In: Proc. 2015 Int. Conf. Comput. Sci. Comput. Intell. CSCI 2015; 2016. p. 362–7,

36. Saeed U, Hamalainen J, Mutafungwa E, Wichman R, Gonzalez D, Garcia-Lozano M. Route-based radio coverage analysis of cellular network deployments for V2N communication. Int Conf Wirel Mob Comput Netw Commun. 2019;2019

37. Wang J, Kwan M-P. Hexagon-based adaptive crystal growth Voronoi diagrams based on weighted planes for service area delimitation. ISPRS Int. J. Geo-Information. 2018;7(7):257.

38. Lu X, Yan H, Li W, Li X, Wu F. An algorithm based on the weighted network Voronoi diagram for point cluster simplification. ISPRS Int J Geo-Information. 2019;8(3):105.

39. Daszykowski M, Walczak B. Density-based clustering methods. In: Comprehensive Chemometrics; 2010.

40. Wong DWS, Huang Q. Sensitivity of DBSCAN in identifying activity zones using online footprints. Proc Spat Accuracy 2016. 2016;

41. Ester M, Kriegel HP, Sander J, Xu X. A density-based algorithm for discovering clusters in large spatial databases with noise. In: Proceedings of the second international conference on knowledge discovery and data mining (KDD'96); 1996. p. 226–31.

42. Zhou C, Frankowski D, Ludford P, Shekhar S, Terveen L. Discovering personal gazetteers. In: Proceedings of the 12th annual ACM international workshop on geographic information systems–GIS '04; 2004. p. 266–73.

Analysis of Weather Effects on People's Daily Activity Patterns Using Mobile Phone GPS Data

7

Abstract

This chapter describes a framework for using a mobile phone GPS data to investigate the effects of weather on people's daily activity routines. Temperature, rainfall, and wind speed are among the weather variables considered in our case study discussed in this chapter. We describe a method for inferring people's daily activity patterns, including the places they visit and when they do so, as well as the duration of the visit, based on GPS position traces of their mobile phones and Yellow Pages information. An analysis of 31,855 mobile phone users reveals that people are more likely to stay at restaurants or food outlets for longer periods of time, and to a lesser extent at retail or shopping sites, when the weather is extremely cold or the ambiance is calm (non-windy). People's activity habits are affected by certain weather conditions when compared to their usual patterns. People's motions and activities are evident at different times of the day. The weather has a wide range of effects on different geographical areas of a large city. When urban infrastructure data is employed to characterize areas, significant connections between weather conditions and people's accessibility to public rail network are observed. This chapter gives a new perspective of how mobile phone GPS data can be utilized in the context of weather's influence on human behavior, specifically choices of daily activities, as well as the impact of environmental factors on urban life dynamics. The conceptual framework and analysis discussed in this chapter are based on the original research by Phithakkitnukoon et al. (PLoS One. 8:12, 2013; PLoS One. 7:10, 2012; Activity-aware map: Identifying human daily activity pattern using mobile phone data. LNCS, 2010).

Keywords

GPS data · Weather effect · Atmospheric condition · Daily activity patterns · Weather variables · Human behavior

7.1   Motivation and State of the Art

People frequently take their mobile phones with them to places because this ubiquitous technology provides the users with a source of constant and available communication as well as personal enjoyment. The accompanying mobile phone, on the other hand, can provide researchers with a useful instrument for recording human mobility behavior. Researchers can potentially gain a better knowledge of both individual and group behaviors that collectively shape our society as a result of this. Telecom companies can track people's phones as they move through the ubiquitous network of towers, in addition to logging incoming and outgoing calls. This turns mobile phones into personal life loggers, providing longitudinal recordings of personal mobility as well as unparalleled fine-grained data at the aggregate level. This can provide researchers with an insight into numerous aspects of human life. For instance, mobile phones can be used to research social structure [1], how a person's social network diversity can lead to greater personal economic development [2], and how weather affects people's use of phone calls to connect with others [3].

There have been several studies on human mobility that used location traces from connected cellular towers, which is beneficial to urban planning and traffic engineering (e.g., [4–11]). Human movement has been studied and characterized in several ways. Human trajectories, for example, exhibit a high degree of temporal and spatial regularity, with a high possibility of returning to a few frequently visited areas [4]. Trajectories of human mobility are guided by the principle of exploration and preferential return, which governs how people travel to new places while frequently returning to places they've already visited [5]. Others attempt to predict individual mobility by combining data from phone location traces (i.e., phone movement) with datasets containing geographical features like point of interest (POI) and land-use information [6]. Despite the diversity of people's travel patterns, there is a high degree of regularity in their mobility on a daily basis, with 93% of people's whereabouts predictable [7]. Developing an understanding of mobility patterns through phone location trace data has helped with detecting the outbreak of mobile phone viruses [8], comparing people flow between cities [9], identifying commuting patterns [10], and understanding the geography of social networks [11].

These emergent studies of mobility have mostly concentrated upon modeling, predicting, and evaluating human mobility data between cities. However, these techniques typically miss out on the fuller context of mobility, such as the type of activities that people might be engaged with at the sites they travel to. After all, individuals migrate between places in the city for varied goals. Besides travelling between home and work, people are also involved in activities connected to the site they visit, for example, eating in a restaurant, shopping or browsing in a mall, and jogging in a park. As a result, creating ways to assist us infer the types of activities connected with various public locations can provide a deeper characterization of people's daily activity patterns, which has several potential benefits, including facilitating urban design and management.

In this chapter, we discussed a framework that describes how to utilize detailed location records from mobile phones and spatial profiles to characterize human daily activity patterns. The analysis described in this chapter builds on and extends a previous investigation that shows how weather conditions can affect people's mobile social interaction [3], which shows how we can learn more about people's behavior in their daily activities by seeking correlations with detailed weather information. After all, studies have demonstrated that weather has an impact on people's mood [12, 13], thermal comfort level [14, 15], and social interaction [3]. Weather can also have an impact on traffic demand and how we travel [16–18], public health [19], crime rates [20], and even stock values [21, 22]. As a result, this chapter discusses our research into how weather influences people's mobility patterns and activities in the Tokyo Metropolis, Japan. For the rest of the chapter, we'll refer to this case study location as Tokyo.

7.2   Methodology

Cities we live in are increasingly associated with unprecedented amounts of data being produced and captured. Here we discuss how to analyze opportunistic datasets used in our analysis to reveal relationships and hidden patterns of inhabitants in a large city like Tokyo. Doing so enables us to provide knowledge that will help to inform better urban design and planning are proactive to the needs of its citizens.

7.2.1   Datasets

Three datasets were used in this case study. The first dataset was GPS position traces of mobile phone users in Tokyo. The data was collected for a full calendar year, from August 1, 2010, to July 31, 2011, during which time each mobile phone user's position was continuously logged. The accelerometer was utilized to detect periods of relative stasis during which power-hungry GPS collection capabilities may be turned off to save battery life. As a result, the sample rate varied according to the user's mobility. However, sampling did not occur more than once every five minutes.

This mobile phone GPS dataset was provided by one of the largest mobile phone operators in Japan. The data was gathered from mobile phone users who had signed up for location-based services. As shown in Fig. 7.1, the location data was sent through the network and used to perform specific analysis, after which certain services were provided to the registered users. Users of mobile phones were made aware that their positions were being captured as part of this program. Furthermore, the company completely anonymized the dataset to protect users' privacy. A unique user ID, position (latitude, longitude), timestamp, altitude, and approximated error (i.e., <100 m (less than 100 m), <200 m, or <300 m) were all included in each entry in the dataset. This dataset provides finer granular location traces than standard mobile phone call detail records (CDRs), which only record the user's location when

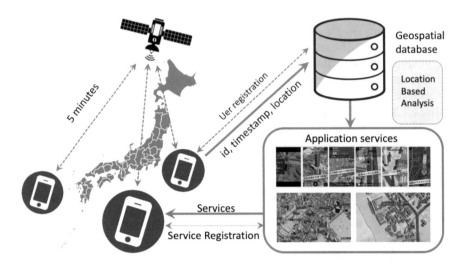

Fig. 7.1 Data collecting process overview

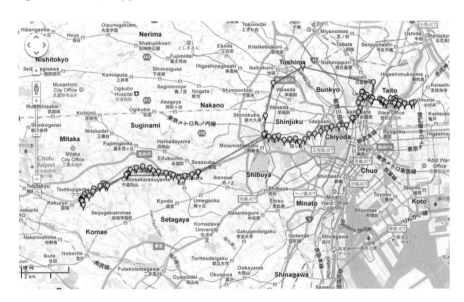

Fig. 7.2 A user's location traces on a mobile phone

a cellular network connection is made, such as when making/receiving a call or sending/receiving a text message. The location traces of a mobile phone user as an example is illustrated in Fig. 7.2.

The weather condition information in Tokyo is the second dataset. Metbroker [23] was used to gather this data. MetBroker provides users access to twelve databases containing data from seven different nations. It is primarily utilized to

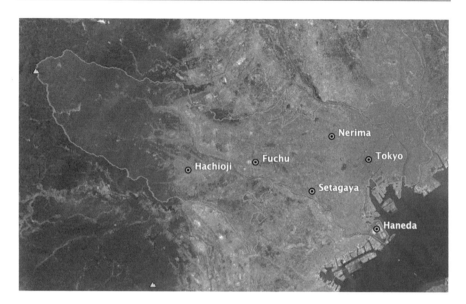

Fig. 7.3 Locations of weather stations from which the data was gathered for the study. The area considered in this study is enclosed by a highlighted contour line

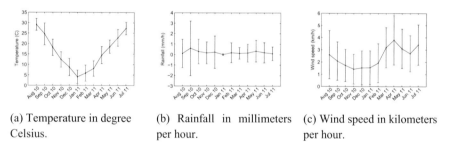

(a) Temperature in degree Celsius.

(b) Rainfall in millimeters per hour.

(c) Wind speed in kilometers per hour.

Fig. 7.4 The monthly weather conditions in Tokyo over the course of the study. (**a**) Temperature in degree Celsius. (**b**) Rainfall in millimeters per hour. (**c**) Wind speed in kilometers per hour

provide critical input for agricultural models, but the data is also relevant to our inquiry. MetBroker is a legacy weather database that provides a uniform format for smooth integration of sensor networks from various weather stations. As a result, it is a trustworthy data source for researchers. Hourly data on Tokyo's temperature (degree Celsius), rainfall (millimeter per hour), and wind speed (kilometers per hour) was collected from six different weather stations between August 1, 2010 and July 31, 2011, their geographical locations shown in Fig. 7.3. Figure 7.4 depicts the monthly statistical means and standard deviations of each meteorological parameter over the course of the study.

The national phone directory or Yellow Pages data collected from Telepoint [24] is the third dataset used in this analysis. Every 2 months, the information on about

Table 7.1 Space profiles categories and corresponding activities

Space profile category	Examples of inferred activities
Eateries	Consuming food and/or beverages
Retail shops/malls	Window shopping, shopping, leisure browsing
Parks/Rivers	Leisure
Financial institutions/employment	Financial transactions
Public transportation	Catching trains or buses
Nightlife retail/entertainment	Pubs/bars activities or nightlife entertainment
Religious institutions	Religious related activities
Educational institutions	Education related activities

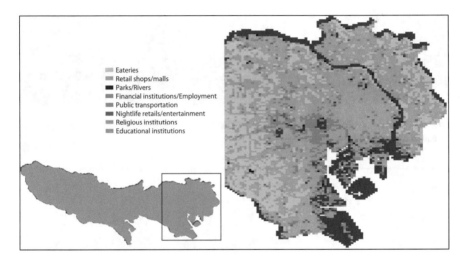

Legend:
- Eateries
- Retail shops/malls
- Parks/Rivers
- Financial institutions/Employment
- Public transportation
- Nightlife retails/entertainment
- Religious institutions
- Educational institutions

Fig. 7.5 A partial representation of the Tokyo map, showing the most likely activities based on yellow-pages information for each grid cell

28 million addresses across the country is geocoded (latitude, longitude). The data from October 2010 was used because it was the most recent database update for the study's time period. These addresses were categorized into 14 different groups. However, only eight of the categories linked with activities that people engage in inside these 52 localities were used. Other categories included residential areas, vacant land, agricultural fields, and so on. This was accomplished using the same method we used to create the Activity-Aware Map [25] in our prior work. This entails utilizing the Weight-Area method to categorize each 250 m by 250 m grid cell, in which each cell is allocated the most probable activity, i.e., the most dominating activity in the cell, based on the space profile category. The (most common) space profile categories and their associated activities are shown in Table 7.1. Based on the Yellow Pages data, a map was created that depicts the most likely activities in various cell regions. A partial picture of Tokyo's activity map, with distinct colored activities is depicted in Fig. 7.5.

7.2.2 Analysis

A mobile phone has become a modern-day necessity and a vital component of our daily life. This work takes use of its widespread use to collect people's everyday activity patterns and mobility. The mobility pattern of each mobile phone user, i.e., trajectories of moves and protracted stop, can be inferred from detailed location records of mobile phone users. People are most likely engaged in some activities when they are not commuting and are at stops. They could be eating at a restaurant, shopping at a mall, relaxing in a park, and so on, in addition to being at home or at work. As a result, we presumed that an activity (other than those done at home or at work) was only engaged during a 'stop' in our analysis. We describe our basic methods for extracting *trips* and *stops* in order to segment these traces into individual trajectories so that each individual's daily mobility pattern can be determined.

Suppose that X denotes a set of sequential traces of the user such that $X = \{x(1), x(2), x(3), \ldots, x(i), \ldots\}$ where $x(i)$ is the ith location of the user. A 'stop' can be defined as a series of locations where the user stays in a particular area for an extensive period of time. As each position i contains location information (*lat, long*) and timestamp (t), i.e., $x(i) = (lat(i), lon(i), t(i))$, a stop is thus regarded as a sequence of positions $\{x(k), x(k + 1), x(k + 2), \ldots, x(k + m)\}$ where the distance between any positions is less than the spatial threshold S_{th} i.e., *distance* $(x(i), x(j)) < S_{th}$ for $i, j \in \{k, k + 1, k + 2, \ldots, k + m\}$ and time spent within the location is greater than the time threshold T_{th}, i.e., $t(k + m) - t(k) > T_{th}$. The position $x(k)$ hence appears as the last position of the previous trip while $x(k + m)$ appears as the first position of the next trip.

After the 'stops' have been detected, each user's home and work locations can be estimated based on the most common stop throughout the night (10 pm–6 am) and day (9 am–5 pm), respectively. The result of the estimated home locations was comparable ($R^2 \approx 0.8$) with the area population density information from the 2006 census data [26] (as shown in Fig. 7.6), and the average computed commuting distance of 24.34 km based on the estimated home and work locations was reasonably close to the average commuting distance of 26 km according to the census data [26]. We were able to gather 31,855 subjects (from the dataset provided to us) whose home and workplace were located within the Tokyo metropolis using the aforesaid method.

After identifying stops in addition to home and work locations for each subject, we were then able to infer people's daily activity pattern, which was described as a succession of most likely engaged activities over the course of a day. The daily activity pattern for each subject was created using a method similar to that used in a prior study [25] that used cell tower-level location traces of mobile phone users. However, because the data in this analysis was more granular, we were able to better determine the most likely activity for each hour (as shown in Fig. 7.7). We were able to examine the data for every hour of the day, inferring the major activity during the hour based on the series of stops and types of visited destinations for each subject, rather than using three-hour time windows as in [25]. We then investigated the impact of the weather on people's mobility and daily activity patterns using the

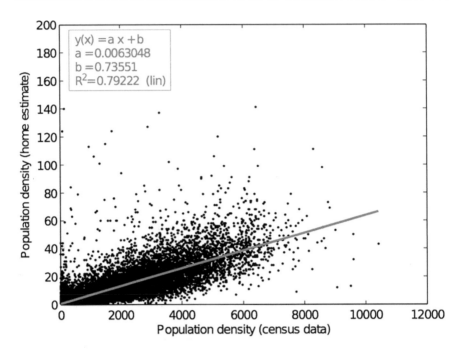

Fig. 7.6 Comparison between our inferred residential municipality population of the mobile phone user subjects and the actual municipality population obtained from census data

Fig. 7.7 Daily activity pattern was captured using an hourly time window

inferred daily activity patterns. Temperature, rainfall, and wind speed were among the weather variables considered. Our analysis looked at the impact of each of these weather variables on people's mobility in terms of stop duration and daily activity patterns throughout time and space.

7.3　Results

Although several weather elements are interdependent, we explored the effects of each weather parameter on people's mobility and activity patterns independently in this study. The following results show how different weather variables affect people's mobility and activity patterns throughout time and space.

7.3.1   Weather Effects on Mobility and Stop Duration

Traveling is done for a variety of purposes. It might be for routine purposeful activities like going to work or shopping, as well as leisurely activities like dining out, socializing, or going on vacation. Understanding the collective mobilities that make up urban dynamics is crucial for municipal governments since it may help with transportation planning and logistics. We decided to investigate how the weather affects people's mobility because the weather has been discovered to have a substantial impact on a number of phenomena connected to human behavior. As a result, we looked at the statistical distribution of stop length for various meteorological variables. We grouped each weather-related characteristic into a series of ranges, or bands, to make it easier to spot any emerging patterns. Temperatures were examined between 25 °C and 35 °C, separated into four bands, each with a 10-degree spread (25 °C to 5 °C, 5 °C to 15 °C, 15 °C to 25 °C, and 25 °C to 35 °C). Rainfall was separated into four bands: no rain (rainfall = 0 mm), the rest was evenly divided into three bands in the range from 0 mm to 15 mm (0–5 mm, 5 mm–10 mm, 10 mm–15 mm); windspeed in four bands: 0–2 kmph, 2–4 kmph, 4–6 kmph, and stronger than 6 kmph. We'd also like to point out that, while other researchers have frequently used the number of visited locations to characterize human mobility (e.g., [4, 5, 7, 11]), this approach would not be appropriate in this work because we wanted to compare people's mobility across different bands of weather for each weather parameter, and each band has different lengths of time observatory.

The statistical distribution of the stop duration for each band of each weather parameter was determined in forms of a probability mass function (pmf), which is essentially a normalized histogram on a logarithmic scale, where normalization permits comparisons across multiple bands of each weather parameter. A logarithmic scale was used because of the observed characteristics of statistical distribution of stop duration as shown in Fig. 7.8a. In addition, Fig. 7.8b and c show the pmf of stop duration for home and workplace, respectively.

When the stop duration is two hours or more, the pmf for the temperature band − 5 °C to 5 °C is higher than other bands, as seen in Fig. 7.9. The result is statistically significant with p-value = 3.0616 x 10^{-4} based on Fisher's exact test with the total

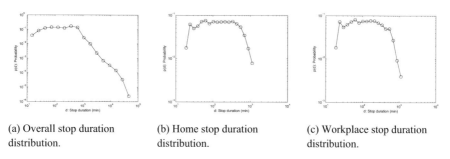

(a) Overall stop duration distribution.

(b) Home stop duration distribution.

(c) Workplace stop duration distribution.

Fig. 7.8   The probability mass function of subject stop duration. (**a**) Overall stop duration distribution. (**b**) Home stop duration distribution. (**c**) Workplace stop duration distribution

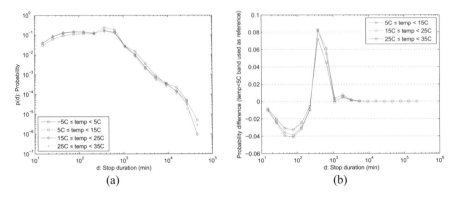

Fig. 7.9 The probability mass function of stop duration for different temperature bands (**a**), as well as the probability difference when compared to the temperature band of less than 5 °C (**b**)

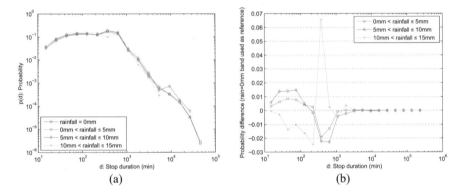

Fig. 7.10 Probability mass function of stop duration under various rainfall bands (**a**) and probability difference relative to the band of rainfall = 0 mm (**b**)

number of stops in very cold weather (temperature < 5 °C) is 262,803, the number of stops in extremely cold weather that are longer than two hours is 188,905, the number of stops in other temperature bands is 2,757,720, and the number of stops in other temperature bands that are longer than two hours is 1,705,941. In other words, on very cold days, there is a significantly higher chance of people stopping for two hours or longer than on other temperature bands. People are more likely to stop for less than two hours on days with temperatures exceeding 5 °C. The difference in probability measures when comparing other temperature bands to the (−5 °C to 5 °C)-band, i.e., subtracting the probability measure of other bands from the (−5 °C to 5 °C)-band is illustrated in Fig. 7.9b. As a result, a positive difference indicates that the (−5 °C to 5 °C)-band has a higher likelihood and vice versa. Rainfall, on the other hand, appears to have no bearing on how long people choose to stop (Fig. 7.10). Wind speed, on the other hand, has a relationship with stop duration, comparable to temperature. For generally calm days with wind speeds between 0–2 kmph, there is a larger possibility (higher pmf) of a two-hour or longer stop

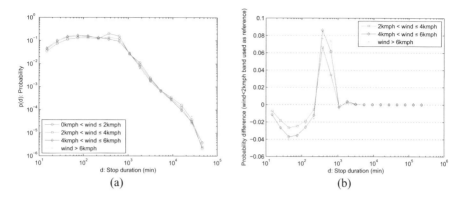

Fig. 7.11 The probability mass function of stop duration for various wind speed bands (**a**), as well as the probability difference when compared to the wind speed band of less than 2 kmph (**b**)

(Fig. 7.11). With the total number of stops in calm weather (wind speed <2 kmph) is 1,855,467, the number of stops in calm weather that are longer than two hours is 1,230,570, the number of stops in other wind-speed bands is 1,489,735, and the number of stops in other wind-speed bands that are longer than two hours is 867,471, the statistical significance test yields p-value of 1.829×10^{-9}. As a result, aside from workplace and home, people rarely take breaks of two hours or more. The only times this does not apply are on extremely cold or quiet days.

Furthermore, using Yellow Pages information and the Activity-Aware Map method, we set out to infer the activities for persons who stopped for two hours or longer. When the temperature was between $-5\,°C$ and $5\,°C$, people spent about 80% of their longer stops at areas that predominantly consist of eateries or food outlets, such as restaurants, cafes, and so on, and 17% at areas that predominantly consist of retailing, such as shops, shopping malls, and so on, after discounting locations identified as home and workplace.

When the wind speed was less than 2 km/h, people spent 88% of their longer stops in areas with eateries and food outlets, and 11% in areas with retail and shopping. These findings suggest that on a chilly or calm (non-windy) day, people are more likely to take their time eating, taking snacks, and/or drinking, and, to a lesser level, spending time on shopping-related activities in retail or in a shopping mall, whether buying or simply window shopping.

7.3.2   Weather Effects on Activities at Different Times of the Day

Human mobility is quite predictable since it occurs on a regular basis [4]. Since most of us are caught up in work-life patterns, our daily activities are frequently pre-scheduled. With a daily routine that most of us find difficult to change, such as going to work in the morning, eating lunch around noon, shopping in the afternoon, and going to a pub in the evening, the question is if the weather has any impact on such daily activity patterns. We explored the weather influence at different times of the day to see if it affects our everyday activities and if so, to what extent. The entropy of activities was utilized to capture the differences in the subjects' activities. Entropy ($H(X)$) is a measure of uncertainty or randomness associated with a random variable in information theory, and it can be computed as follows [27].

$$H(X) = -p(x_i) \sum_i^n \log_2 p(x_i), \tag{7.1}$$

where X is a random variable with n outcomes $\{x_1, x_2, \ldots, x_i, \ldots, x_n\}$ and $p(x_i)$ is a probability of outcome x_i.

The entropy value was calculated for each hour of the day, where X represents a collection of activities participated in by each of n subjects in that hour, i.e., variable x_i indicates the subject i's activity. The probability $p(x_i)$ was then calculated as a ratio of the number of times the activity x_i was observed across all n subjects in an hour to the total number of subjects (in our case, $n = 31,855$). The entropy was chosen for this analysis because it was appropriate for categorical variables that represented activities. A greater entropy score indicates that the subjects' activities are more random. When entropy is equal to zero, there is no randomness, which means that all subjects' activities are the same.

The results reveal that varied weather conditions indeed have an impact on people's activity patterns throughout the day when the activity pattern is inferred in the same way as explained before for each band of each weather parameter. As shown in Fig. 7.12, entropy values drop dramatically between 8 a.m. and 9 a.m., regardless of the day's temperature range where there is generally low variation in people's activities. Given the timeframe, it's possible that this is due to the fact that most people are on their way to work. The effect of different temperature bands becomes more obvious after 10 a.m. The effect of the temperature range $-5\,°C$ to $5\,°C$ is distinct from that of the other temperature ranges. Entropy values for this band increase starting at 10 a.m. and peak between 2 and 6 p.m. before declining (see Fig. 7.12). By 10 p.m., this band's entropy value is still larger than the other bands. In other words, from 10 a.m., on very cold days ($-5\,°C$ to $5\,°C$), people's activities are most varied.

The reverse effect is observed for the other temperature bands (temperature bands above $5\,°C$). The temperature range of 15 to $25\,°C$ has the least diversified activity, notably between 12 and 5 p.m. In comparison, the temperature range from 5 to $15\,°C$ has slightly higher entropy values, followed by the temperature band from 25 to $35\,°C$. In other words, people's activities vary the least in the three temperature bands above $5\,°C$, with the least variance in the moderately comfortable temperature range of $15\,°C$ to $25\,°C$, and the most variation in the highest temperature range of $25\,°C$ to $35\,°C$. In fact, on days when the temperature is $5\,°C$, people's activities begin to vary after 5 p.m. and continue well into the night. The warmest days ($25\,°C$ to $35\,°C$) appear to have the most fluctuation in nighttime activities. In other words, on very warm evenings, individuals tend to engage in a wide range of activities.

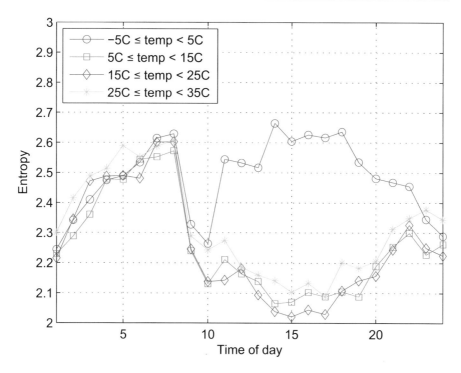

Fig. 7.12 Entropy values at different times of the day and for different temperature bands

Rainfall has a comparable effect to really cold days in terms of causing large fluctuations in people's activities. In fact, the strongest rainfall band (10 mm– 15 mm) has an entropy level that exceeds that of a very cold day. In other words, on days when the rain is the hardest, people's activities are the most diverse.

Similarly, just like that of temperature, the entropy values reveal a significant reduction between 8 and 9 a.m. regardless of the amount of rainfall. After 9 a.m., this pattern changes (see Fig. 7.13). One band stands out from the rest: no rainfall (rainfall = 0 mm). On dry days, from 9 a.m. to 4 p.m., the entropy values of people's activities decrease (i.e., the variation of their activities decreases), with the lowest entropy values occurring between 3 and 4 p.m.

However, rainfall (measured in millimeters) follows the opposite tendency as dry days. On rainy days, we see an increase in entropy values starting at 9 a.m., indicating more variance in people's activities. In fact, the more rainy the day is (i.e., the higher the rainfall), the more diverse people's activities are. For example, we notice that the highest rainfall band (10 mm to 15 mm) has the most diverse activities, especially in the summer between 3 and 7 p.m. Lower rainfall encourages a wider range of activities, though not as much during the day.

During the day, the effect of wind differs little from that of temperature and rainfall. In fact, we've noticed that wind speed has the greatest impact on people's actions. When the wind speed exceeds 4 kmph, there is higher entropy levels (see

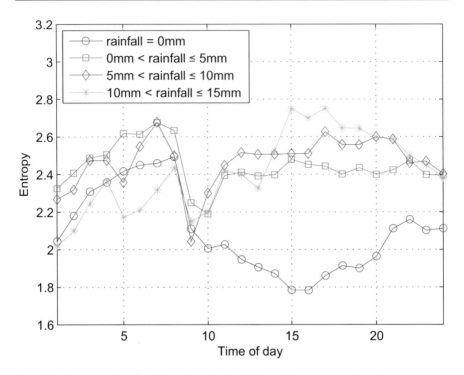

Fig. 7.13 Entropy values for different bands of rainfall at different times of the day

Fig. 7.14). This is especially noticeable in the highest wind speed band (when it is 6 kmph or greater). This has a big impact on the entropy values, especially between 11 a.m. and 10 p.m. People's activities are highly diversified on days with such high wind speeds, especially within that time span. On the other hand, between 8 and 9 a.m. on days with the highest wind speed band, entropy values drop dramatically. As a result, during strong wind mornings, people's variety of activities is highly limited. However, days with relatively low winds (0 to 2 kmph) may not always result in lower entropy values. Days with wind speeds between 2 kmph and 4 kmph have the lowest entropy levels between 1 p.m. and 8 p.m. In other words, people's activities vary the least on days with light winds, notably between 1 and 8 p.m.

7.3.3 Weather Effects on Activities in Different Areas

The impact of weather on people's activities, as well as the extent to which weather influences people's activity choices, may vary depending on where they live in the city. Given the scale of the city of Tokyo, this is very crucial. Shopping malls, hospitals, houses of worship, parks, and other geographical aspects of the metropolitan landscape could be involved. It can also include public transportation networks in large metropolises.

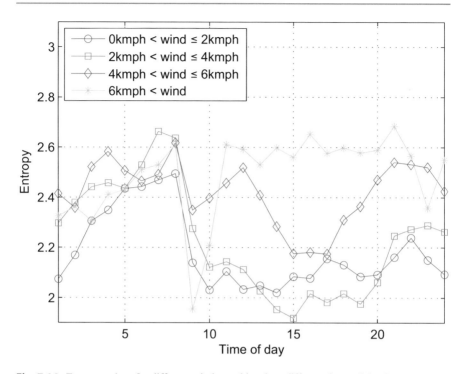

Fig. 7.14 Entropy values for different wind speed bands at different times of the day

The entropy was employed as a measure of variation in activity patterns to see how the weather affects different districts of Tokyo. This time, we calculated the impact of weather as a difference in entropy values between regular and irregular weather for each of Tokyo's 52 municipalities. We defined the regular condition for each weather parameter as 5 °C to 35 °C for temperature, 0 mm for rainfall, and 0 to 4 kmph for wind speed, based on the results from the previous section on temporal weather effects. For temperature, rainfall, and wind speed, the irregular condition are −5 °C-5 °C, >0 mm, and > 4kmph, respectively.

The findings, as depicted in Figs. 7.15, 7.16 and 7.17, show that each weather parameter has a different impact in different parts of Tokyo at different times of the day. The impact of temperature on people's normal activity patterns in various districts of Tokyo is shown in Fig. 7.15. Furthermore, the influence changes depending on the time of day. For example, we can see in Fig. 7.15b that weather impacts significantly upon people living in the western region of Tokyo between 4 a.m. and 7.59 a.m., and also between 10 a.m. and 12.59 p.m., when compared to other regions. However, between 1 and 3.59 p.m., the effect is more evenly distributed throughout Tokyo.

Rain has the greatest impact on people's daily activities between 4 a.m. and 6.15 a.m. (Fig. 7.16b), and then gradually fades between 7 a.m. and 9.59 a.m. (Fig. 7.16c). Windy days provide particularly intriguing patterns, notably in Tokyo's

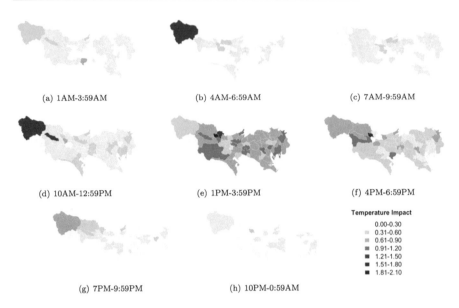

Fig. 7.15 The impact of temperature on daily activity patterns in various municipalities

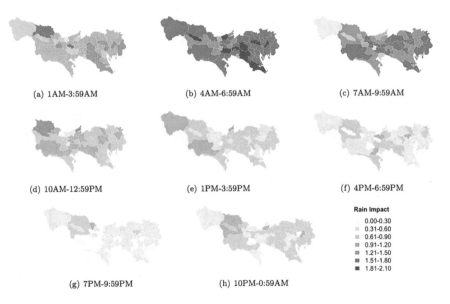

Fig. 7.16 The impact of rain on daily activity patterns across different municipalities

western region (Fig. 7.17). To begin with, while wind has an impact on people's normal activities in most areas of Tokyo, especially between the hours of 4 a.m. and 9.59 a.m., this impact is less pronounced in the western region. However, from

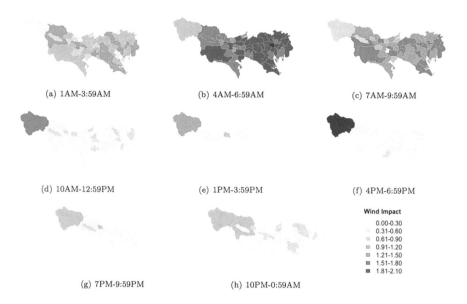

(a) 1AM-3:59AM   (b) 4AM-6:59AM   (c) 7AM-9:59AM

(d) 10AM-12:59PM   (e) 1PM-3:59PM   (f) 4PM-6:59PM

Wind Impact
  0.00-0.30
  0.31-0.60
  0.61-0.90
  0.91-1.20
  1.21-1.50
  1.51-1.80
  1.81-2.10

(g) 7PM-9:59PM   (h) 10PM-0:59AM

Fig. 7.17 The impact of wind on daily activity patterns across different municipalities

10 a.m. onwards, the impact of wind becomes more noticeable (significantly more so than in other parts of Tokyo), peaking between 4 and 6.59 p.m.

A number of influential factors could be responsible for the various impacts described above. One of them is people's capacity to move around in various weather situations. As a result, we decided to investigate these findings further in light of people's access to public transportation. As a measure of public transportation accessibility, we calculated the distance between the subject's home location and the nearest public transportation hub for each municipality. Bus stops and train stations were chosen as public transportation hubs in this analysis as buses and trains are the most common modes of public transportation in Tokyo.

The ability to access trains has been demonstrated to have a significant impact on people's activities in Tokyo [28]. However, the placement of train stations in various districts of Tokyo may not be as dense, and people may be further distant from available rail stations. In fact, in the far west, there is one town with only one train station, and no one in our study population lives within 2 kilometers of it. So we wanted to examine if people's proximity to a railway station influences their activity choices in various weather conditions.

The findings shown in Fig. 7.18 reveal that the farther a person is from a railway station, the greater the influence of that weather pattern on people's activity choices ($R^2 = 0.7 \sim 0.8$). Bus stations, on the other hand, are seldom far away (average distance is 215 m). People's proximity to bus stops appears to show that varying weather patterns have no discernible effect on people's choice of activities when it comes to buses. We did the same analysis for people's proximity to other types of

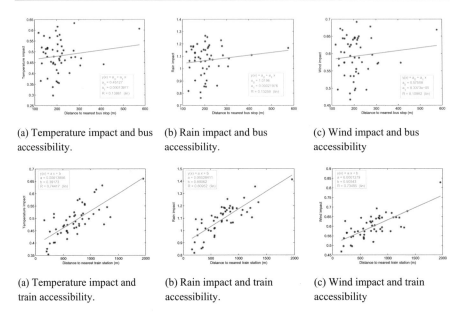

(a) Temperature impact and bus accessibility.

(b) Rain impact and bus accessibility.

(c) Wind impact and bus accessibility

(a) Temperature impact and train accessibility.

(b) Rain impact and train accessibility.

(c) Wind impact and train accessibility

Fig. 7.18 The relationship between weather impact and public transportation accessibility. (**a**) Temperature impact and bus accessibility. (**b**) Rain impact and bus accessibility. (**c**) Wind impact and bus accessibility. (**d**) Temperature impact and train accessibility. (**e**) Rain impact and train accessibility. (**f**) Wind impact and train accessibility

Table 7.2 The R^2 values representing the correlation between the impact of different weather parameters and accessibility of urban infrastructure

	Weather parameters		
Urban infrastructure	Temperature	Rain	Wind
Train station	0.744	0.810	0.735
Bus stop	0.451	0.133	0.110
Hospital	0.223	0.070	0.051
Shopping mall	0.073	0.194	0.142
Park	0.376	0.080	0.215
Night club	0.087	0.055	0.025

urban infrastructure (hospitals, shopping malls, parks, and nightclubs), but there were no significant correlations. The R^2 values is displayed in Table 7.2.

7.4   Conclusion

Researchers can better interpret human behavior and gain insights into various domains through the utilization of the growing availability of big data. In this chapter, we discuss a research study that analyzes GPS trajectories of 31,855 mobile phone users in Tokyo for a full calendar year. On the basis of their location traces and yellow-page information, daily activity patterns of mobile phone users were estimated. With temperature, rainfall, and wind speed used as weather parameters, we were particularly interested in the impact of the weather on people's daily activity patterns. We've discovered fascinating variations in how various weather conditions affect people's mobility and activities in Tokyo.

People were more likely to stay longer and spend more time in areas with eateries and food outlets such as restaurants, cafes, and so on, and to a lesser extent in shopping areas with retail outlets, shopping malls, and so on, on days that were very cold (-5 °C to 5 °C) or calm (i.e., wind speed less than 2 km/h). Furthermore, on really cold days, we discovered that people's activities were more varied during the day, especially after 10 a.m., with the greatest fluctuations occurring between 2 and 6 p.m. Similarly, we discovered that on wet days (particularly between 10 a.m. and midnight), as well as on days when the wind speed was greater than 4 km/h (mean $= 2.6$ km/h), people's activities were more diverse. Finally, distinct weather influences were detected for different geographical locations. The farthest western section of Tokyo appears to be the most affected by extreme cold, with residents' routine activities being considerably disrupted in the early mornings and before midday. We discovered considerable correlations between the impact of weather and local residents' accessibility to railway stations by categorizing diverse places according to people's accessibility to urban infrastructure i.e., distance to the nearest accessible point.

The findings we provide in this study do, however, have certain limitations. There could be minor variances in meteorological conditions in locations not near the weather stations studied, which could influence the findings. Even when dealing with the effects of weather on such a high level and on such a wide scale, how individuals truly feel about the weather is subjective and remains an unresolved question. One of the promising ways is participatory mobile sensing [29]. Although there is interdependence between meteorological parameters and how people perceive the overall weather situation, this intricacy was not taken into account in the study. Future research can take this into account in order to have a better grasp of the weather's overall impact. Important social events like New Year's and Christmas, as well as emergency situations like earthquakes, were not taken into account in this study, which could have influenced the results. Social events, emergency situations, and day type are thus worth exploring further in the future. Finally, due to the limited availability of GPS data and the low resolution of the spatial profile characterization utilized in this investigation, the activity patterns inferred from location traces and yellow-pages information may not accurately reflect actual activities. Nonetheless, we believe that the results of this study, to a considerable extent, constitute fresh understanding regarding the impact of weather on our behavior, particularly our physical activity patterns.

As the worldwide trend toward urbanization continues, the capacity to combine large-scale datasets to discover specific patterns in human behavior in large metropolises will become more significant. With more people coming to cities and as cities grow, one of the key issues facing urban planners is the ability to deliver a wide range of services in an effective and efficient manner while also ensuring the safety of its residents. This chapter shows how we may start to leverage big data to provide richer contexts into people's behavior and activities in order to improve our understanding of citizens in a big city. We might imagine how such information can guide a wide range of decisions by developing methods that can help us infer not only mobility but also where people stop and what they do (apart from being at home and at work) under different weather conditions. On a more immediate level, this covers public transportation schedules, future transportation hub design, retail/restaurant staffing levels, and service opening times, among other things. Knowing where people concentrate in light of weather patterns can also be used to predict how security and emergency services can be best deployed. Of course, such capacity to obtain live models and create predictive models of inhabitants in large metropolises can be greatly improved through greater access to better quality and higher resolution data, and by refining our approaches as well as computational efforts. The approach discussed in this chapter also paves the way for future efforts in which researchers combine multiple sources of data to help mitigate some of the issues that plague rapidly growing metropolises around the world, such as waste management, pollution control, and even crime reduction by determining where police deployment is most useful and likely to be effective.

References

1. Onnela J-P, et al. Structure and tie strengths in mobile communication networks. Proc Natl Acad Sci. 2006;104(18):7332–6.
2. Eagle N, Macy M, Claxton R. Network diversity and economic development. Science (80-.). May 2010;328(5981):1029–31.
3. Phithakkitnukoon S, Leong TW, Smoreda Z, Olivier P. Weather effects on Mobile social interactions: a case study of Mobile phone users in Lisbon, Portugal. PLoS One. 2012;7:10.
4. González MC, Hidalgo CA, Barabási AL. Understanding individual human mobility patterns. Nature. 2008;453:779–82.
5. Song C, Koren T, Wang P, Barabási AL. Modelling the scaling properties of human mobility. Nat Phys. 2010;6:818–23.
6. Calabrese F, Di Lorenzo G, Ratti C. Human mobility prediction based on individual and collective geographical preferences. In: 13th international IEEE conference on intelligent transportation systems; 2010. p. 312–7.
7. Song C, Qu Z, Blumm N, Barabási AL. Limits of predictability in human mobility. Science (80-.). 2010;327(5968):1018–21.
8. Wang P, González MC, Hidalgo CA, Barabási A-L. Understanding the spreading patterns of Mobile phone viruses. Science (80-.). 2009;324(5930):1071–6.
9. Isaacman S, Becker R, Cáceres R, Kobourov S, Rowland J, Varshavsky A. A tale of two cities. In: Proceedings of the eleventh workshop on Mobile computing systems & applications–HotMobile '10; 2010. p. 19–24.
10. S. Isaacman et al., "Identifying important places in People's lives from cellular network data," 2011.
11. Phithakkitnukoon S, Smoreda Z, Olivier P. Socio-geography of human mobility: a study using longitudinal mobile phone data. PLoS One. 2012;7(6):e39253.
12. Charry JM, Hawkinshire FB. Effects of atmospheric electricity on some substrates of disordered social behavior. J Pers Soc Psychol. 1981;41(1):185–97.
13. Howarth E, Hoffman MS. A multidimensional approach to the relationship between mood and weather. Br J Psychol. 1984;75(1):15–23.
14. Trenberth KE, Miller K, Mearns L, Rhodes S. Effects of changing climate on weather and human activities. J Chem Educ. 2002;79(4):433.

15. Tucker P, Gilliland J. The effect of season and weather on physical activity: a systematic review. Public Health. 2007;121(12):909–22.
16. Cools M, Moons E, Creemers L, Wets G. Changes in travel behavior in response to weather conditions. Transp Res Rec J Transp Res Board. 2010;2157(1):22–8.
17. Maze TH, Agarwal M, Burchett G. Whether weather matters to traffic demand, traffic safety, and traffic operations and flow. Transp Res Rec J Transp Res Board. 2006;1948(1):170–6.
18. Agarwal M, Maze TH, Souleyrette RR. Impacts of weather on urban freeway traffic flow characteristics and facility capacity. In: Proceedings of the 2005 mid-continent transportation research symposium; 2005. p. 14.
19. Patz JA, Engelberg D, Last J. The effects of changing weather on public health. Annu Rev Public Health. 2000;21(1):271–307.
20. Cohn EG. Weather and crime. Br J Criminol. 1990;30(1):51–64.
21. Saunders E Jr. Stock prices and the wall street weather. Am Econ Rev. 1993;83(5):1337–45.
22. Akhtari M. Reassessment of the weather effect: stock prices and wall street weather. Michigan J Bus. 2011;7(1):51–70.
23. National Agricultural Research Center. Metbroker. 2012.
24. Zenrin: Maps to the Future. Telepoint Pack!. 2011.
25. Phithakkitnukoon S, Horanont T, Di Lorenzo G, Shibasaki R, Ratti C. Activity-aware map: identifying human daily activity pattern using mobile phone data, vol. 6219. LNCS; 2010.
26. Statistics Bureau. 2006 establishment and Enterprise census. Ministry of Internal and Communications. 2006;
27. Shannon CE. A mathematical theory of communication. Bell Syst Tech J. 1948;27 (July):379–423.
28. Tokyo Metropolitan Region Transportation Planning Commission, "Tokyo metropolitan region-person trip survey," 1998.
29. Weddar. How does it feel? Weddar; 2012.

Analysis of Tourist Behavior Using Mobile Phone GPS Data

8

Abstract

This chapter discusses a framework for analyzing tourist behavior that employs a large-scale opportunistic mobile sensing approach through a case study of mobile phone users in Japan. It deliberates on how enormous mobile phone GPS location records can be utilized in analyzing tourist travel behavior through a number of proxies, including the number of trips made, time spent at places, and mode of transportation used. Furthermore, this chapter examines the relationship between personal mobility and tourist travel behavior, and it reveals a number of meaningful insights for tourism management, including tourist flows, top tourist destinations and origins, top destination types, top modes of transportation in terms of time spent and distance traveled, and how personal mobility information can be used to estimate a likelihood in tourist travel behavior, such as the number of trips initiated, time spent at destination, and trip distance. In addition, the chapter describes an application developed based on the findings of this analysis allowing its user to monitor touristic, non-touristic, and commuting trips, as well as home and work locations and tourist flows, which can be useful for city planners, transportation managers, and tourism authorities. This chapter is inspired by our original work by (Phithakkitnukoon et al., Pervasive Mob Comput 18: 2015) and Horanont et al. (Lecture notes of the institute for computer sciences, social-informatics and telecommunications engineering, LNICST: 2015).

Keywords

GPS data · Tourist behavior · Touristic trip inference · Travel behavior · Personal mobility · Touristic flow analysis

8.1   Motivation and State of the Art

Cities have grown increasingly instrumented and networked as a result of recent breakthroughs in information and communications technology (ICT). Sensors integrated in urban systems (e.g., CCTV, building access systems, public Wi-Fi) as well as personal electronic gadgets are continually measuring and recording the activities and movements of city people (e.g., mobile phones, laptops, tablets). Individual digital traces are generated in enormous numbers, allowing community and city-level behavioral signatures to be collected. An image of urbanization from the real (physical) world can be digitally rebuilt as a whole. As a result, as described in recent studies, an examination of the characteristics of a city, its functionalities, and the behavior of its inhabitants can be undertaken. For example, based on an analysis of connected cell tower locations (Call Detail Records or CDRs) of nearly one million mobile phone users, Phithakkitnukoon et al. [1] introduced the concept of a map that describes most likely activities in different areas of a city, from which it was found that people who work in the same industry (e.g., restaurant, retail, etc.) tend to have similar daily activity patterns. Gonzalez et al. [2] discovered that, despite the diversity of our travel histories, human people follow simple repeating patterns. Similarly, Song et al. [3] discovered that people's movements are 93% predictable. Later, Song et al. [4] developed a model that captures people's habit of traveling between fixed sites on a regular basis. Phithakkitnukoon et al. [5] contributed to these efforts by demonstrating that people's travel patterns are influenced by the geography of their social ties.

One of the most pressing ecological and social issues of the twenty-first century is human mobility. People travel for a variety of reasons, including commuting and tourism. Commuting journeys are frequently repeated with the same routes, making them somewhat predictable. Touristic trips, on the other hand, are less predictable. As a result, it is critical for urban planners, transportation managers, and tourism authorities to understand tourist travel behavior. Today's ubiquitous technologies, such as mobile phones, which have become an indispensable part of many people's lives, and which, as evidenced by recent research studies, have sensing capabilities that enable the phone to function as a personal sensor, collectively create a new sensing paradigm that includes humans as part of a sensing infrastructure. Researchers are able to obtain an unprecedented amount of fine-grained behavioral data from people by utilizing the sensor capabilities of mobile phones. It has a number of advantages over traditional tourist behavior surveys.

Surveys and questionnaires are commonly used in traditional tourist behavior studies. For example, Alegre and Pou [6] studied the length of stay on the Balearic Islands in Spain using survey data collected from 56,915 tourists over 3 years. To evaluate the length of stay on holidays in Bodrum, Turkey, Gokovali et al. [7] analyzed three-week questionnaire data obtained from 1023 tourists. Wu et al. [8] investigated the decision-making process of Japanese visitors using survey data from 1253 respondents in Japan.

As we have seen in several studies discussed in the previous chapters, we once again capitalized on the large-scale opportunistic mobile sensing approach to examine tourist behavior in this chapter. We examined 130,861 mobile phone users' GPS location traces over the course of a year in Japan. This chapter will detail our strategy for detecting tourists using GPS location information, which allows us to undertake tourist behavior research and illustrate applications for urban planners, transportation managers, and tourism authorities using these records. Three primary contents of this chapter includes a large-scale (country-level) analysis of tourist behavior, including tourist flows, time spent at destinations, mode of transportation, relationship between personal mobility and travel behavior, and similarity in travel behavior; a computational framework for identifying touristic trips from GPS location information, including algorithm for home and work location detection; and a prototype application developed based on the analysis that allows the user to observe and analyze touristic trips and flows.

8.2 Methodology

8.2.1 Dataset

By capitalizing on the opportunistic sensing paradigm in which mobile phones are used as personal sensors for location tracking, GPS location records from 130,861 mobile phone users in Japan over the course of a calendar year (1 January 2012–31 December 2012) were used for our analysis. The information was provided to us by one of Japan's largest mobile phone companies, and it was gathered from users who signed up for location-based services (and given consent for the use of their location information). As shown in Fig. 8.1a, the location data was sent through the network and used to perform specific analysis, for which certain services were provided to the registered users. The dataset was entirely anonymized by the mobile phone operator before being sent to us in order to protect users' privacy. A unique user ID, position (latitude, longitude), timestamp, altitude, and approximate error (i.e., 100, 200, or 300 m) were all included in each entry in the dataset. An accelerometer was utilized to determine periods of relative stillness during which

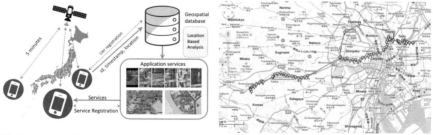

(a) Overview of data collection process.   (b) An example of a mobile phone user's location traces.

Fig. 8.1 Data collection process (**a**) and an example data (**b**). (**a**) Overview of data collection process. (**b**) An example of a mobile phone user's location traces

power-consuming GPS collection tasks may be paused, reducing battery consumption. As a result, the sample rate fluctuated depending on the user's mobility, although it never exceeded once every five minutes. Figure 8.1b displays the location traces of a mobile phone user in our dataset as an example.

For various reasons, such as the phone being turned off, not being subscribed to a service, or travel abroad, some of the subjects in our pool of subjects did not have GPS location traces for the entire year of our study; therefore, to ensure a sufficient amount of data for our analysis, we selected the 130,861 subjects whose GPS locations were observed at least 350 days out of 365 days in our data.

8.2.2 Residence and Workplace Location Detection

To identify tourists from GPS trajectories, we must first determine the subjects' home and workplace locations in order to extract and focus only on non-commuting trips. Touristic journeys can be determined from such non-commuting excursions. The method involves three steps, as shown in Fig. 8.2, to detect home and workplace locations.

The initial phase was to locate a *stop*, which was a collection of nearby GPS sites where the user spent a significant amount of time. A stop could be a person's home, workplace, restaurant, or market, for example. In order to gather ground truth, we recruited 15 individuals to carry a smartphone for a month and use an application that allowed them to track their daily stops. We discovered that the most accurate spatial and temporal criteria [9] for identifying stop were 196 m and 14 min, as shown in our experimental results in Figs. 8.3a and b. If $X_u = \{x_{t_1}, x_{t_2}, \ldots, x_{t_i}, \ldots\}$

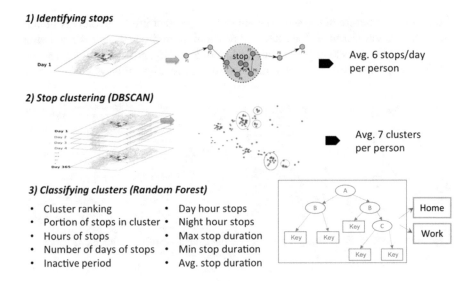

Fig. 8.2 Home and workplace location detection method

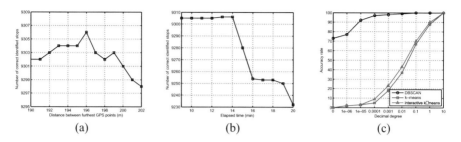

Fig. 8.3 Experimental results for stop criteria in terms of space and time

denotes a set of GPS locations of user u where x_{t_i} is the location at time t_i, then our experimental results suggested that we group $x_{t_i}, x_{t_{i+1}}, x_{t_{i+2}}, \ldots, x_{t_m}$ that are within 196 m and $t_m - t_i \leq 14$ min as a stop.

The next step was a spatial clustering of the detected stops. The cluster's centroid was regarded as a noteworthy location (e.g., home, workplace, other). We tested a number of clustering algorithms and discovered that DBSCAN (density-based spatial clustering of applications with noise) [10] performed the best among them. We created a tool for validation that allowed the tool user to label significant locations after observing clusters of stops, as illustrated in Fig. 8.4. We used this tool to tag our data with the locations of 400 individuals' homes and workplaces, which were used as ground truth in our validation. The DBSCAN method outperformed k-means and interactive k-means in locating the centroids of stop clusters that corresponded to locations of significant places, according to our experimental data (see Fig. 8.3c). DBSCAN with $\varepsilon = 30$ m and MinPts $= 5$ points achieved nearly 100% accuracy rate at 0.0001 decimal units (about 11.1 m), whereas k-means and interactive k-means can only achieve less than 10%. Note that decimal degree of 1.0 is approximately 111.32 km.

The final step was to categorize significant locations as either home or workplace. For this task, we used hand-labeling ground truth and found that the Random Forest [11] was the best performing method when compared to the k-nearest neighbors (k-NN) and Naïve Bayesian classifiers (see Table 8.1, ten-fold cross-validation was used) using the following 10 features:

- **Cluster ranking**: high-ranking clusters may indicate where people live and work.
- **Portion of stops in cluster**: People tend to visit essential areas, such as home and workplace, more frequently than other places, so this suggests the value of places to some level.
- **Hours of stops**: this is the time period during which clustered stops appeared. This feature would be 8/24 if stops were observed from 9 a.m. to 4 p.m. (during the year).
- **Days with clustered stops**: the number of days with clustered stops.
- **Inactive hours**: for each subject, an inactive period was defined as a three-hour period in which the number of GPS locations was less than the norm. A fraction

(a) Hand labeling tool.

(b) Labeling example. Green dots are GPS locations, yellow dots are stops, and purple circle is cluster centroid.

Fig. 8.4 A screenshot of our developed hand labeling tool and an example of labelling. (**a**) Hand labeling tool. (**b**) Labeling example. Green dots are GPS locations, yellow dots are stops, and purple circle is cluster centroid.

of clustered stops that fall into the inactive period are represented by the inactive-hours attribute.

- **Day-hour stops**: portion day hours (10 p.m.–6 a.m.) in which stops were observed.
- **Night-hour stops**: portion night hours (9 a.m.–6 p.m.) in which stops were observed.
- **Maximum stop duration**: the maximum duration of a stop.
- **Minimum stop duration**: the minimum duration of a stop.
- **Average stop duration**: the average duration of a stop.

We compared our results with the census data to further validate our home location estimation and found that the estimated population density based on our home location estimation was comparable ($R^2 = 0.966$) to city population density information obtained from the 2006 Japanese Census [12], as shown in Fig. 8.5.

Table 8.1 Performance comparison for classification of home and workplace

Classifier	Home		Workplace		Other		Total	
	Precision	Recall	Precision	Recall	Precision	Recall	Precision	Recall
Random Forest	87.32	87.57	89.94	77.28	94.19	98.58	90.48	87.18
k-NN	76.82	84.32	71.56	56.54	87.70	91.50	78.69	77.45
Naïve Bayes	71.93	84.91	78.75	71.36	96.35	94.50	82.34	83.59

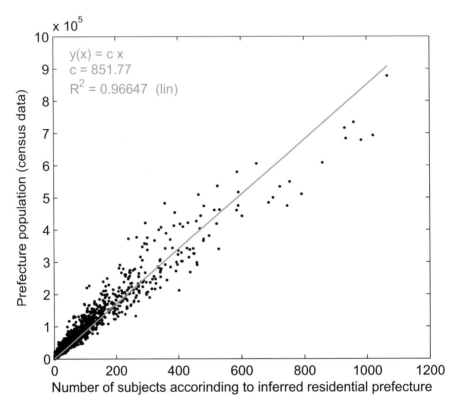

Fig. 8.5 The difference between our estimated residential city population of mobile phone users and the actual city population based on census data

8.2.3   Touristic Trip Inference

We were able to distinguish between commuting (between home and workplace) and non-commuting trips after acquiring a firm estimate of home and workplace locations. Suppose that $S_u = \{s_{1, j}, s_{2, j}, \ldots, s_{i, j}, \ldots\}$ is a collection of user u's stops, where $s_{i, j}$ is the ith stop with spatial profile j; a trip is a collection of stops that starts at home and finishes at home. To put it another way, $trip = \{s_{m + 1, home}, s_{m + 2, j}, s_{m + 3, j}, \ldots, s_{m + t, home}\}$, where t is the total number of stops in the trip. As a result, a commuting trip is defined as one with at least one stop at a workplace $s_{i, work}$, i.e., $commuting = \{s_{m + 1, home}, s_{m + 2, j}, \ldots, s_{m + 3, work}, \ldots, s_{m + t, home}\}$. A non-commuting trip, on the other hand, is defined as one in which none of the stops are at a workplace, i.e., $non\text{-}commuting = \{s_{m + 1, home}, s_{m + 2, j}, \ldots, s_{m + 3, j}, \ldots, s_{m + t, home}\}$.

We defined touristic trip as non-commuting journey with at least one stop at a touristic destination, i.e., $touristic\ trip = \{s_{m + 1, home}, s_{m + 2, j}, \ldots, s_{m + 3, touristic}, \ldots, s_{m + t, home}\}$. The information on touristic attractions published by the Ministry of Land, Infrastructure, and Transport of Japan (MLIT) [13] was used in our analysis.

The touristic destination information is made up of two sets of data: one contains destination locations (latitude and longitude), and the other contains destination area in the forms of polygon (i.e., GIS shapefiles). These two datasets do not overlap; in other words, they do not contain the same destination locations. Both sets of data were utilized to classify stops as touristic or non-touristic. A touristic stop is defined as one that is located within 200 m of a touristic destination or within the polygon area covered by a touristic destination. As a result, a spatial profile j can be classified as residential, commercial, touristic, or non-touristic.

8.3 Analysis of Tourist Behavior

The detected tourist trips were then used in our tourist behavior analysis. The number of subjects from 47 different Japanese prefectures based on our home location estimation is shown in Fig. 8.6. The spatial distribution of these subjects is illustrated in Fig. 8.7. The corresponding prefecture names are listed in Table 8.2. Tokyo has the highest number of subjects (15,586) followed by Kanagawa (8588) and Saitama (7329) as expected (because our subject distribution correlates well with the census population density (Fig. 8.5).

In our tourist behavior analysis, we were interested in trip flows—the number of tourist trips made to and from different prefectures in Japan, time spent at destination, modes of transportation used by the tourists, and correlations between personal mobility and touristic travel behavior. In great part the environmental and social effects of tourism are a product of the sheer volume of tourist trips [14]. Therefore, it is important to understand these characteristics of tourist flows in order to make better informed decisions as we move toward sustainable tourism.

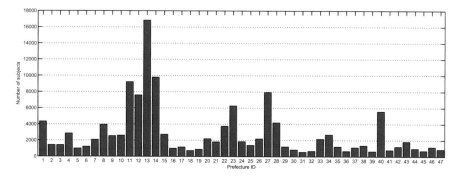

Fig. 8.6 The number of subjects from various prefectures

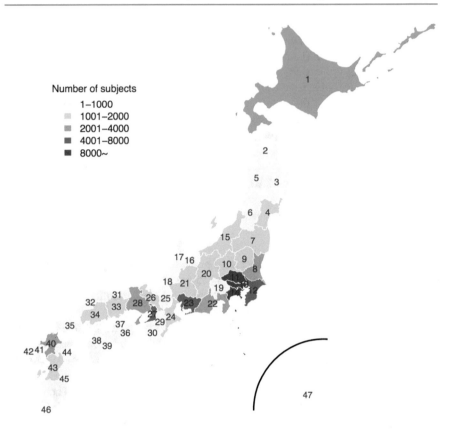

Fig. 8.7 Spatial distribution of the subjects

8.3.1   Amount of Touristic Trips

Touristic trips were extracted for each subject using the method described previously. Figure 8.8a depicts the distribution of the subjects' total number of trips. A total of 1,987,858 trips were made, equating to 15.19 excursions per person on average. Short (less than 5 km), medium (5–100 km), and long (more than 100 km) trips were all categorized (farther than 100 km). Figure 8.8b depicts the distribution of the number of trips in each of these three ranges.

Figure 8.9a illustrates the (sorted) outgoing touristic trips from different prefectures based on the estimated home locations of the subjects, with Tokyo having the highest outflow volume at 495,904, followed by Kanagawa (161,422), and Fukuoka (109,962), respectively. The percentage of trips taken that correspond to trip distance (short, medium, and long) is depicted in Fig. 8.9b.

In all prefectures, medium trips seemed to make up the majority of the travels. Similarly, if the visited prefectures are considered destinations, incoming flows are shown in Fig. 8.10a, with Tokyo as the top destination (1,754,902), Kyoto

Table 8.2 Prefecture identifications and corresponding names

Prefecture ID	Prefecture name	Prefecture ID	Prefecture name
1	Hokkaido	25	Shiga
2	Aomori	26	Kyoto
3	Iwate	27	Osaka
4	Miyagi	28	Hyogo
5	Akita	29	Nara
6	Yamagata	30	Wakayama
7	Fukushima	31	Tottori
8	Ibaraki	32	Shimane
9	Tochigi	33	Okayama
10	Gunma	34	Hiroshima
11	Saitama	35	Yamaguchi
12	Chiba	36	Tokushima
13	Tokyo	37	Kagawa
14	Kanagawa	38	Ehime
15	Niigata	39	Kochi
16	Toyama	40	Fukuoka
17	Ishikawa	41	Saga
18	Fukui	42	Nagasaki
19	Yamanashi	43	Kumamoto
20	Nagano	44	Oita
21	Gifu	45	Miyazaki
22	Shizuoka	46	Kagoshima
23	Aichi	47	Okinawa
24	Mie		

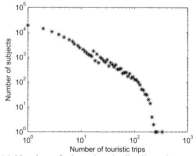

(a) Number of trips taken is distributed.

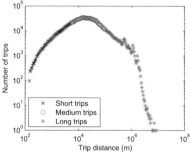

(b) Trip distance distribution for several ranges: short (less than 5 km), medium (5-100 km), and long (greater than 100 km).

Fig. 8.8 The distribution of trip length and number of trips. (**a**) Number of trips taken is distributed. (**b**) Trip distance distribution for several ranges: short (less than 5 km), medium (5–100 km), and long (greater than 100 km)

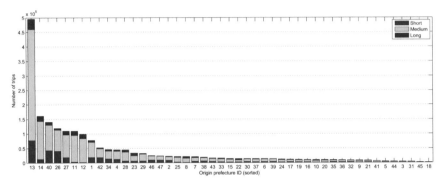

(a) Total number of outbound journeys from various prefectures (sorted).

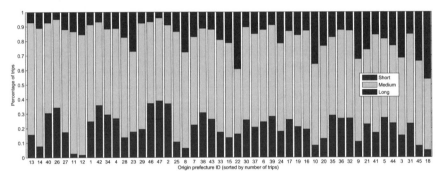

(b) Percentage of outbound journeys from various prefectures (short, medium, and long) (sorted by total number of trips).

Fig. 8.9 Trips taken from various prefectures. (**a**) Total number of outbound journeys from various prefectures (sorted). (**b**) Percentage of outbound journeys from various prefectures (short, medium, and long) (sorted by total number of trips)

(341,911), and Fukuoka (278,886). The majority of the trips for the inflow volumes were a mix of short and medium-range trips, as shown in Fig. 8.10b.

The Origin–Destination matrix (O–D matrix), which defines trip distribution and flow, is one of the most essential components examined in transportation management and engineering. The O–D matrix can be derived from touristic trip origin and destination information in our case, and it is shown in Fig. 8.11 where we can see the top origin and destination prefectures as well as the trip distribution.

The number of travels to and from the same prefectures is relatively large, as can be seen in the O–D matrix, which is intuitive. Within the same region, a high number of trips can be seen between nearby prefectures, such as the Northeast region (including Hokkaido and Tohoku, prefecture ID 1–7), the Kanto region (prefecture ID 8–14), the Kansai region (prefecture ID 24–30), and the West region (including Okinawa and Kyushu, prefecture ID 40–46). The O–D matrix also reveals that prefecture population density appears to be related to outflow volume but not inflow

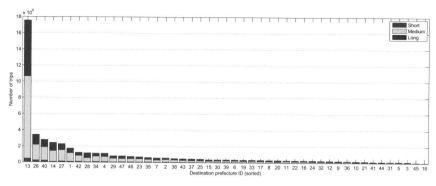

(a) The total number of incoming trips in each prefecture (sorted).

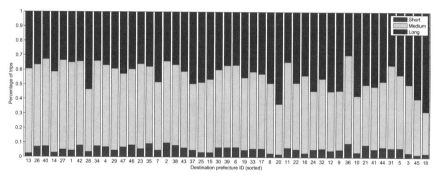

(b) Incoming travel percentages in different prefectures (short, medium, and long) (sorted by total number of trips).

Fig. 8.10 Trips taken from various prefectures. (**a**) The total number of incoming trips in each prefecture (sorted). (**b**) Incoming travel percentages in different prefectures (short, medium, and long) (sorted by total number of trips)

volume; for example, the number of trips made from densely populated areas like the Greater Tokyo region (Saitama, prefecture ID 11), Chiba (prefecture ID 12), Tokyo (prefecture ID 13), and Kanagawa (prefecture ID 14) is significantly higher than other prefectures, but the inflow volume is significantly lower. Tokyo and Kanagawa, on the other hand, have substantially higher inflow volumes due to the numerous tourist attractions in respective prefectures. Tokyo is Japan's capital city; Kanagawa's capital city is Yokohama, which features Japan's and the world's largest Chinatowns; and Kamakura, another Kanagawa city, is known for Buddhist temples and Shinto shrines. Our findings suggest that Kanagawa draws a large number of visitors from Tokyo. Other instances include Osaka and Kyoto, two prefectures in close location that create opposite-direction flows; being a more densely populated prefecture, Osaka (2,666,371 inhabitants) has a far higher outflow than Kyoto (2,666,371 inhabitants) (1,474,473 inhabitants). Kyoto, on the other hand, receives far more visitors than Osaka due to its principal tourist attractions,

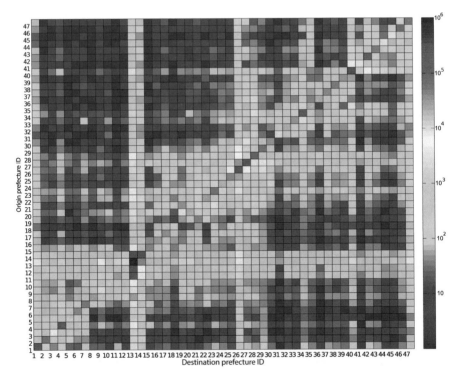

Fig. 8.11 Origin–Destination matrix (number of tourism visits detected to/from various prefectures)

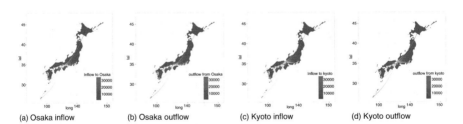

Fig. 8.12 Osaka and Kyoto's tourist influx and outflow. (**a**) Osaka inflow (**b**) Osaka outflow (**c**) Kyoto inflow (**d**) Kyoto outflow

which include several World Heritage sites. Fig. 8.12 depicts the inflows and outflows of Osaka and Kyoto. The flow between Yamaguchi (prefecture ID 35) and Fukuoka (prefecture ID 40), which are prefectures near to each other separated by the Kanmon Straits (a stretch of sea that separates Japan's main islands), is another intriguing finding from the O-D matrix. Yamaguchi is about an hour's train ride from Hakata in Fukuoka; Yamaguchi has a lovely town that prospered as "Kyoto in the West" during the medieval period, and it is one of the region's top tourist destinations. The O–D matrix shows enormous fluxes between the two prefectures, with a flow from Fukuoka to Yamaguchi that is slightly bigger than the flow in the opposite direction.

8.3.2 Time Spent at Destination

Another essential feature of tourist behavior is the amount of time travellers spend at their destination. We estimated the time spent at destination simply as the total amount from the time of arrival at the first destination until the departure time of the last destination of a trip using the extracted touristic trips. Tourists can visit more than one destination in a single trip.

As a result, all destinations on a single trip were considered the overall destination where the tourist spent time. The average time spent on a trip was 543.40 min (about 9 h), and the standard deviation was 764.39 min (about 12.5 h), as shown in Fig. 8.13a. Figure 8.13b shows the distribution of the likelihood of time spent at destination (normalized by the total number of trips) in short, medium, and long trips for different lengths of trips; from this figure, it can be seen that tourists are more likely to spend more than 546.83 min (about 9 h) on longer trips.

To investigate geography and time spent by tourists, Fig. 8.14a displays the average time spent at destinations by tourists from various origin prefectures. When compared to tourists from other prefectures, Niigata tourists spend the greatest time at places on a trip (733.66 min or about 12 h), followed by Nagasaki (696.47 min or about 11.5 h) and Shimane (664.69 min or about 11 h). Figure 8.14b depicts the average time spent in various destination prefectures. Nagasaki and Hokkaido revealed to be the top prefectures where tourists spent the greatest time on their destinations (327.17 min and 326.31 min (about 5.5 h), respectively) followed by Tokyo (312.82 min or about 5 h).

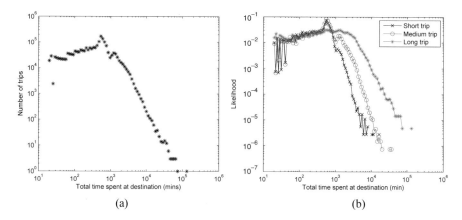

(a)   (b)

Fig. 8.13 Distribution of time spent on tourist trips and the likelihood of doing so. (**a**) Distribution of the amount of time spent on a trip. (**b**) Likelihood of amount of time spent on a short, medium, and long trips

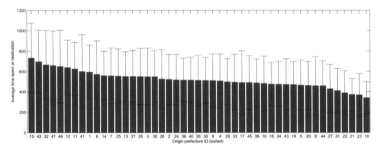

(a) Average time spent at destination(s) on a trip by tourists from different prefectures (sorted).

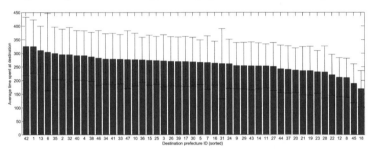

(b) Average time spent at destination prefectures (sorted).

Fig. 8.14 Average time spent at destinations by tourists from and to different prefectures. (**a**) Average time spent at destination (s) on a trip by tourists from different prefectures (sorted). (**b**) Average time spent at destination prefectures (sorted)

If there is any doubt about the differences between the results in Figs. 8.14a and b, we would like to point out that the average time spent at destinations by tourists from different prefectures (Fig. 8.14a) is higher than the average time spent at different destination prefectures (Fig. 8.14b) because tourists from different origin prefectures can visit more than than one destination, which makes the total time spent on different destinations on their single trips higher than the time spent at particular destinations (in different prefectures), which is shown in Fig. 8.14b.

Tourists may be attracted to different types of destinations. There were 26 touristic destination types examined in our analysis, based on information provided by Japan's Ministry of Land, Infrastructure, and Transport [13]. Table 8.3 displays a list of these different tourist destination types. These 26 touristic destination classifications were assigned to all of the destinations found in our data. Figure 8.15 depicts the number of journeys made to various types of destinations based on this information. The most popular types of tourist destinations were shrines and temples, buildings, and annual events. Figure 8.16 depicts the average amount of time spent at various types of destinations; as can be seen, Open field was the most popular (445.89 min or about 7 h) followed by Botanical garden and aquarium (378.51 min or about 6 h) and Museum (311.78 min or about 5 h).

Table 8.3 Touristic destination types

Destination type ID	Type
1	Botanical garden and aquarium
2	Zoo
3	Museum
4	Open field
5	Historic site
6	National landscape
7	Castle
8	Mountain
9	Cave
10	Cape
11	Canyon
12	Island
13	Annual event
14	Garden and park
15	Building
16	Vegetation
17	Historic landscape
18	River
19	Coast
20	Lake
21	Wetland
22	Waterfall
23	Shrine and temple
24	Shrine garden and park
25	Natural phenomena observatory
26	Plateau

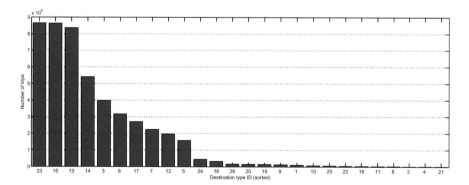

Fig. 8.15 Number of trips made to different types of destinations (sorted)

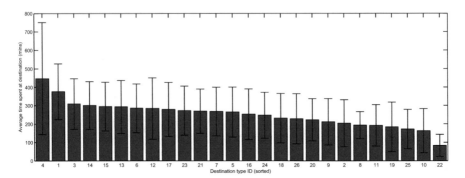

Fig. 8.16 Average time spent at different types of destinations (sorted), with standard deviation bars

8.3.3 Mode of Transportation

Following our investigation of the number of visits to acquire a better knowledge of trip distribution and flow, as well as time spent at locations by area and type, we expanded our research to look into another significant component of tourism: the mode of transportation used by tourists.

We used the framework we previously established [15] for identifying types of transportation used by mobile phone users based on their GPS locations in our analysis of this research. Our previous framework essentially looked at the GPS traces along with detected stops, i.e., $\{s_{1,\ j}, x_t, x_{t\ +\ 1}, x_{t\ +\ 2}, \ldots, x_{t\ +\ m}, s_{2,\ j}, \ldots\}$ and defined a *segment* as a series of GPS locations between adjacent stops, i.e., $\{x_t, x_{t\ +\ 1}, x_{t\ +\ 2}, \ldots, x_{t\ +\ m}\}$. Based on the rate of change in velocity and train line proximity, these segments were then categorized as walking or non-walking. The walking segments were inferred as walking when used as a mode of transportation. Non-walking segments are then classified into one of two modes of transportation: car or train, using the Random Forest classification technique and the following features: segment distance, time elapsed in segment, speed (minimum, maximum, average, maximum acceleration, and rate of change in velocity), and portion of GPS points that fall into train and road networks. The framework was tested against the ground truth modes of transportation used by 100 subjects who were asked to carry a smartphone with a mobile app that allowed them to input the modes of transportation they used on a daily basis for a month. With an overall precision rate of 86.89% and a recall rate of 84.17%, this framework performed admirably.

Walking, driving, and taking the train were therefore the modes of transportation considered in this analysis. Tourists can travel using a variety of forms of transportation in a single journey. As a result, we examined the percentage of time spent on various forms of transportation during a tourist excursion, as well as the percentage of distance traveled by various modes. Figure 8.17a depicts the total distribution of time spent on various forms of transportation during a single trip, from which we can see that on average tourists spend somewhat more time in a car (48.57%) than

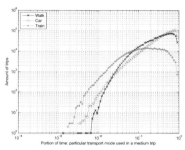

(a) Overall distribution of time portion spent on different modes of transportation.

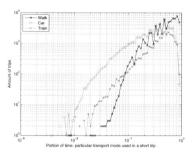

(b) Portion of time spent on different modes of transportation in a short trip.

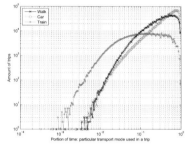

(c) Portion of time spent on different modes of transportation in a medium trip.

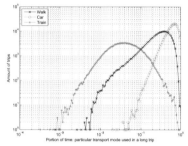

(d) Portion of time spent on different modes of transportation in a long trip.

Fig. 8.17 Distribution of portion of time spent by train, car, and walking throughout a trip. (**a**) Overall distribution of time portion spent on different modes of transportation (**b**) Portion of time spent on different modes of transportation in a short trip. (**c**) Portion of time spent on different modes of transportation in a medium trip. (**d**) Portion of time spent on different modes of transportation in a long trip

walking (45.21%), but significantly more time in a car and walking than on a train (6.22%). Tourists (on average) spent more time walking (78.98%) than driving (14.89%) or taking the train (6.14%) on short trips, as shown in Fig. 8.17b. On average, tourists were more likely to spend longer time in a car (53.87%) than walking (39.35%) or using a train (6.78%) on medium journeys as shown in Fig. 8.17c. Tourists on long trips spent more time on average in cars (64.04%) than walking (33.11%) or taking the train (2.85%), as shown in the distributions in Fig. 8.17d. Table 8.4 lists the mean and standard deviation values for the results in Fig. 8.17. Overall, when a trip lengthens, more time is spent in a car but less time is spent walking; by contrast, time spent in a train stays the shortest of all forms of transportation.

Distribution of the portion of a trip's distance covered by various means of transportation is depicted in Fig. 8.18a, from which we discovered that the average percentage of distance traveled by car (71.51%) was larger than the percentage traveled by walking (2.52%) or by train (6.97%). Tourists traveled by automobile 72.46% of the time for short excursions (Fig. 8.18b), compared to walking (20.42%)

Table 8.4 Means and standard deviations of percentage of time on a trip spent on different transport modes (Fig. 8.17)

| | Mode of transportation | | | | | |
| | Walking | | Car | | Train | |
Trip type	Mean	Std.	Mean	Std.	Mean	Std.
Short	78.98	35.50	14.89	30.36	6.14	20.56
Medium	39.35	27.91	53.87	29.10	6.78	15.57
Long	33.11	17.66	64.04	17.86	2.85	4.90
Overall	45.21	32.19	48.57	32.17	6.22	15.76

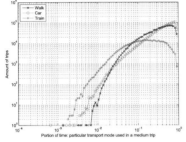

(a) Overall distribution of distance travelled by different modes of transportation.

(b) Portion of trip distance travelled by different modes of transportation in a short trip.

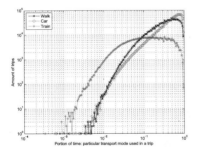
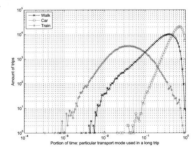

(c) Portion of trip distance travelled by different modes of transportation medium trip.

(d) Portion of trip distance travelled by different modes of transportation in a long trip.

Fig. 8.18 Distribution of portion of trip distance travelled by train, car, and walking throughout a trip. (**a**) Overall distribution of distance travelled by different modes of transportation. (**b**) Portion of trip distance travelled by different modes of transportation in a short trip. (**c**) Portion of trip distance travelled by different modes of transportation in a medium trip. (**d**) Portion of trip distance travelled by different modes of transportation in a long trip

and train (8.12%). Tourists traveled by vehicle (79.50%) for greater distances than they did by walking (12.67%) or rail (8.83%) for medium journeys (Fig. 8.18c). Similarly, tourists preferred to go larger distances by automobile (95.33%) rather than walking (3.27%) or taking the train (1.40%). Overall, vehicles are used for

Table 8.5 Means and standard deviations of percentage of time distance on different transport modes (Fig. 8.18)

| | Mode of transportation | | | | | |
| | Walking | | Car | | Train | |
Trip type	Mean	Std.	Mean	Std.	Mean	Std.
Short	72.46	40.61	20.42	35.90	7.12	22.67
Medium	12.67	18.38	79.50	25.15	7.83	17.76
Long	3.27	4.02	95.33	5.79	1.40	3.47
Overall	21.52	32.32	71.51	34.89	6.97	17.82

travel on excursions more frequently than trains and walking; as trips increase longer, cars are used to travel longer distances, while walking lengths decrease; trains, on the other hand, remain the least used method of transportation among others. Table 8.5 show the means and standard deviations.

8.3.4 Relationship Between Personal Mobility and Travel Behavior

Given knowledge of visitors' personal mobility patterns, can we anticipate or estimate tourist travel behavior, such as travel frequency, journey distance, time spent at locations, and type of destination more likely to be visited? In our study, we wanted to find an answer to this question.

Personal mobility was investigated utilizing three matrices: *mobility level*, *travel dispersion*, and *travel scope*. The total number of stops, which measures activity or how far a person travels, was used to determine a person's mobility level. To determine how spread out or dispersed a person's mobility is, we used travel dispersion, which can be calculated as $\sqrt{\sigma_{lat}^2 + \sigma_{lon}^2}$, where σ_{lat} and σ_{lon} are the standard deviation of points along latitude and longitude, respectively. When computing standard deviation, i.e., $\sigma = \sqrt{\frac{1}{N}\sum_{i=1}^{N}(x_i - \bar{x})^2}$, the quantity $(x_i - \bar{x})$, i.e., distance from the mean was computed with harvesting formula with kilometers as the unit of measurement. In other words, the magnitude of the resultant vector of the standard deviation of latitude and longitude can be interpreted as the journey dispersion. The last criterion used to measure personal mobility was travel scope, which was simply defined as a person's most distant destination. In other words, the farthest destination was the GPS location point that was the furthest away from the person's home location. Since we only investigated destinations in Japan for this study, the farthest points studied were limited to within Japan.

We first investigated the link between mobility and travel behavior, which was measured by the number of trips taken, the distance traveled, and the time spent at the destination. Figure 8.19 depicts the results for the mobility level. The touristic and non-touristic mobility levels (ML) were evaluated. The results demonstrate a strong association between touristic ML and the average number of travels, with $r^2 = 0.902$ ($\alpha = 0.247$ for the linear regression of the form $y = \alpha x$), which is logical; in other words, a higher degree of touristic mobility indicates that a person is more

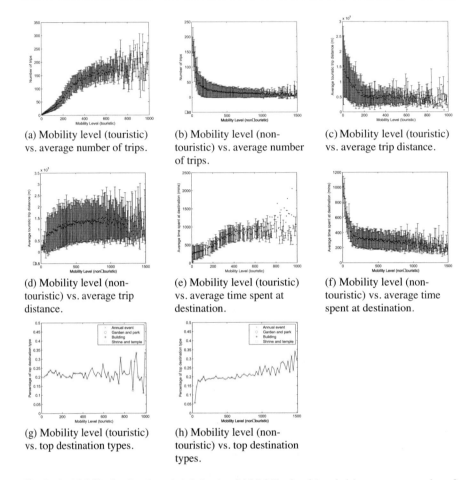

(a) Mobility level (touristic) vs. average number of trips.

(b) Mobility level (non-touristic) vs. average number of trips.

(c) Mobility level (touristic) vs. average trip distance.

(d) Mobility level (non-touristic) vs. average trip distance.

(e) Mobility level (touristic) vs. average time spent at destination.

(f) Mobility level (non-touristic) vs. average time spent at destination.

(g) Mobility level (touristic) vs. top destination types.

(h) Mobility level (non-touristic) vs. top destination types.

Fig. 8.19 Mobility level and touristic behavior. (**a**) Mobility level (touristic) vs. average number of trips. (**b**) Mobility level (non-touristic) vs. average number of trips. (**c**) Mobility level (touristic) vs. average trip distance. (**d**) Mobility level (non-touristic) vs. average trip distance. (**e**) Mobility level (touristic) vs. average time spent at destination. (**f**) Mobility level (non-touristic) vs. average time spent at destination. (**g**) Mobility level (touristic) vs. top destination types. (**h**) Mobility level (non-touristic) vs. top destination types

likely to take more trips. The results demonstrate a substantial inverse association between the number of trips and the non-touristic mobility level for ML > 200 ($r^2 = 0.745$, $\alpha = 0.012$) and ML ≤ 200 (r^2 0.878, $\alpha = 0.348$). When looking at the link between tourism mobility level and trip distance, the results demonstrate a strong inverse association when ML at most 200 stops (r^2 0.953, $\alpha = 574.33$), implying that the higher the touristic ML, the shorter the trip distance. The greater the tourist ML, the longer the time spent at the destination ($r2 = 0.852$, $\alpha = 1.418$); in other words, the higher the tourist ML, the longer the time spent at the destination (on average). Non-touristic ML, on the other hand, was found to have a strong

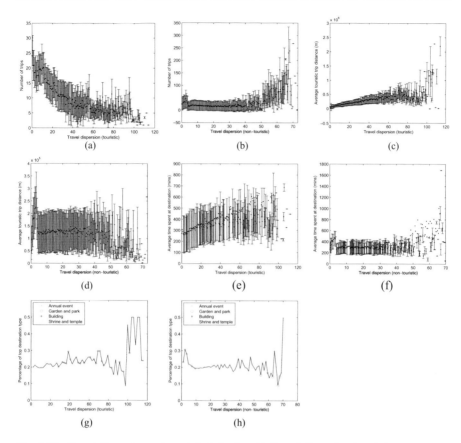

Fig. 8.20 Travel dispersion and touristic behavior. (**a**) Travel dispersion (touristic) vs. average number of trips. (**b**) Travel dispersion (non-touristic) vs. average number of trips. (**c**) Travel dispersion (touristic) vs. average trip distance. (**d**) Travel dispersion (non-touristic) vs. average trip distance. (**e**) Travel dispersion (touristic) vs. average time spent at destination. (**f**) Travel dispersion (non-touristic) vs. average time spent at destination (**g**) Travel dispersion (touristic) vs. top destination types. (**h**) Travel dispersion (non-touristic) vs. top destination types

inverse association with time spent at destination when $ML > 200$ ($r^2 = 0.735$, $\alpha = 0.249$) and when $ML \leq 200$ ($r^2 = 0.883$, $\alpha = 3.126$), implying that the greater the non-touristic ML, the less time spent at destination. The top destination types and their corresponding percentages among other types are shown in Fig. 8.19g and h for different levels of touristic and non-touristic ML. Building was the most popular form of touristic ML site, followed by Annual Event, Shrine, and Temple at all levels of ML. Shrine and Temple were the most popular destinations for non-touristic ML, followed by Annual Event and Building. When ML reached around 300 stops, Shrine and Temple became a clear dominant kind.

For travel dispersion (TD), as shown in Fig. 8.20, there is a strong inverse relationship between touristic TD and the number of trips ($r^2 = 0.862$, $\alpha = 0.077$),

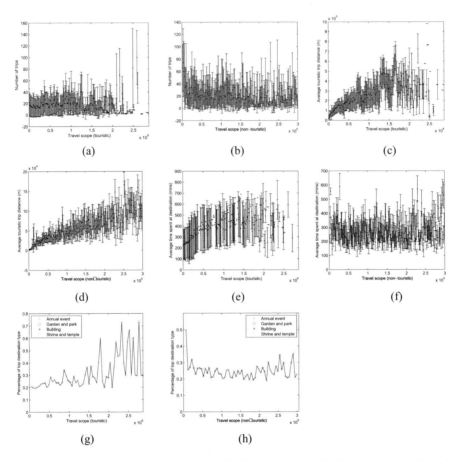

Fig. 8.21 Travel scope and touristic behavior. (**a**) Travel scope (touristic) vs. average number of trips. (**b**) Travel scope (non-touristic) vs. average number of trips. (**c**) Travel scope (touristic) vs. average trip distance. (**d**) Travel scope (non-touristic) vs. average trip distance. (**e**) Travel scope (touristic) vs. average time spent at destination. (**f**) Travel scope (non-touristic) vs. average time spent at destination. (**g**) Travel scope (touristic) vs. top destination types. (**h**) Travel scope (non-touristic) vs. top destination types

implying that the higher the touristic TD (i.e., the more spread out the touristic destinations are), the fewer the trips. We also discovered a substantial link between touristic TD and trip distance ($r^2 = 0.957$, $\alpha = 6457.6$) when touristic TD $\leq$ 80 km. When tourist TD $\leq$ 22 km, Shrine and Temple was found to be the top destination type followed by Annual Event but when TD > 22 km. Annual Event was observed to be the top type followed by Building.

For travel scope (TS), as shown in Fig. 8.21, we observed a strong relationship between non-touristic TS and trip distance ($r^2 = 0.866$, $\leq$0.391), which interestingly suggests that larger non-touristic TS implies a longer trip distance. In other words, the result tells us that people who visit places located farther away for non-touristic purposes are more likely to make longer touristic trips. For destination types, Annual Event appeared to be the top type overall followed by Building. Shrine and Temple was observed as top destination when TS < 500 km. For non-touristic TS, Building was the top destination type followed by Shrine and Temple, and Annual Event. All values of r^2 and linear regression constant α of the results shown in Figs. 8.19, 8.20, and 8.21 are listed in Table 8.6.

8.4   Analysis of Similarity in Travel Behavior

Furthermore, we investigated a cluster of people based on their travel behavior in a smaller scale analysis of travel behavior. Gaining a knowledge of people's travel behavior and their clusters within a prefecture, i.e., having an insight into where people share common activity and what that behavior is, is both intriguing and vital for urban planning.

For each of the 4,734,878 total trips made, we used the trip distance, total time spent at destinations, time spent walking, distance traveled by walking, time spent traveling by car, distance traveled by car, time spent traveling on a train, and distance traveled by train as features to represent the characteristics of travel behavior. We used the k-means clustering technique with $k = 4$ to cluster the trips based on these features for each prefecture. For this preliminary cluster observation, the number of clusters k was chosen arbitrarily. Due to the large number of data instances (4,734,878), other clustering techniques that do not need pre-assignment of the number of clusters, such as DBSCAN, are not suitable.

Following the clustering of all trips, each subject was assigned to a cluster using a majority vote procedure based on the subject's clustered trips. If there was a tie, the subject was assigned randomly to one of the clusters of the subject's clustered trips. We discovered that a large number of subjects (more than 70%) in most prefectures were clustered together, implying that most people have similar travel behavior within the prefecture, based on the clustering results. Nonetheless, other prefectures have a less skewly clustered population. The percentage of people in the prefecture who were assigned to the top cluster (i.e., the most populated cluster) is shown in Fig. 8.22.

The clustering results did not clearly show that people's travel behavior was strongly influenced by their place of residence on a geographical level. Subjects' home locations of each cluster are shown in Figs. 8.23 and 8.24, respectively, for Saitama (prefecture ID 11, where nearly 90% of people's travel behavior is similar) and Fukui (prefecture ID 18, where less than 60% of people's travel behavior is similar). People who travel in similar ways appear to live in different parts of the prefecture. However, future research should look at the impact of the residential region on people's travel behavior, such as examining detailed characteristics of the area in terms of transportation accessibility or urban infrastructure.

Another fascinating aspect of the similarities in travel behavior is the O–D matrix, which describes the characteristics of trips taken from one prefecture to another and even inside the same prefecture (Fig. 8.11). Using the same set of features as before,

Table 8.6 R-squared values and linear regression constant α of the results shown in Figs. 8.19, 8.20, and 8.21 (ML = mobility level, TD = travel dispersion, TS = travel scope)

Personal mobility	Tourist travel behavior proxies					
	Touristic			Non-touristic		
	Number of trips	Trip distance	Time spent at destination	Number of trips	Trip distance	Time spent at destination
ML	$r^2 = 0.902$ $\alpha = 0.247$	$r^2 = -0.629$ $\alpha = -71.683$ For ML > 200, $r^2 = -0.238$ $\alpha = -67.467$ For ML ≤ 200, $r^2 = -0.953$ $\alpha = -574.33$	$r^2 = 0.852$ $\alpha = 1.418$	$r^2 = -0.562$ $\alpha = -0.013$ For ML > 200, $r^2 = -0.745$ $\alpha = -0.012$ For ML ≤ 200, $r^2 = -0.878$ $\alpha = -0.348$	$r^2 = 0.443$ $\alpha = 126.93$	$r^2 = -0.657$ $\alpha = 0.255$ For ML > 200, $r^2 = -0.735$ $\alpha = -0.249$ For ML ≤ 200, $r^2 = -0.883$ $\alpha = -3.126$
TD	$r^2 = -0.862$ $\alpha = -0.077$	$r^2 = 0.632\alpha = -71.683$ For TD > 80, $r^2 = 0.444$ $\alpha = 6105.7$ For TD ≤ 80, $r^2 = 0.957$ $\alpha = 6457.6$	$r^2 = 0.545$ $\alpha = 5.58$	$r^2 = 0.582$ $\alpha = 0.582$	$r^2 = -0.544$ $\alpha = 2085.1$	$r^2 = 0.448$ $\alpha = 9.018$
TS	$r^2 = -0.0048$ $\alpha = 0.00001$	$r^2 = 0.581$ $\alpha = 0.192$	$r^2 = 0.505$ $\alpha = 0.00028$	$r^2 = -0.290$ $\alpha = 0.0001$	$r^2 = 0.866$ $\alpha = 0.391$	$r^2 = 0.074$ $\alpha = 0.0014$

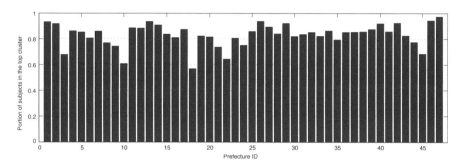

Fig. 8.22 Portion of the number of subjects assigned to the top cluster in each prefecture

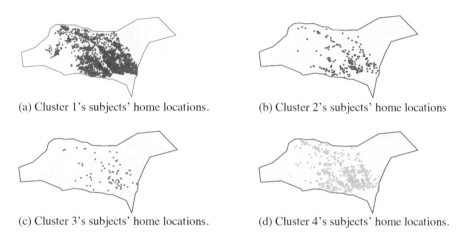

(a) Cluster 1's subjects' home locations.   (b) Cluster 2's subjects' home locations

(c) Cluster 3's subjects' home locations.   (d) Cluster 4's subjects' home locations.

Fig. 8.23 Subjects' homes in different clusters in Saitama. (**a**) Cluster 1's subjects' home locations. (**b**) Cluster 2's subjects' home locations (**c**) Cluster 3's subjects' home locations. (**d**) Cluster 4's subjects' home locations

we calculated the similarity of trips in the O–D matrix by first normalizing each feature (separately) to [0, 1], then multiplying by the number of features. The standard deviation for each normalized feature i that is S_{nor}^i, from which the trip similarity can be calculated as $1 - mean\left(\left\{S_{nor}^1, \ S_{nor}^2, \ \ldots, \ S_{nor}^8\right\}\right)$.

Figure 8.25 illustrates the O–D similarity matrix, which reveals that journeys within the same prefecture are often more similar than trips to and from another prefecture. It's also worth noting that the O–D similarity matrix resembles the O–D matrix (shown in Fig. 8.11). The correlation coefficient (a measure of linear correlation) between the O–D matrix and the O–D similarity matrix was calculated to be 0.6769, indicating that the number of trips and trip similarity had some relationship. In other words, as a destination becomes more popular (i.e., attracts a greater number of visits (or trips), visitors (both new and repeat) tend to travel to the destination in a similar manner—presumably following suggestions from other people who have previously visited the destination (for new visitors) or repeating the same traveling

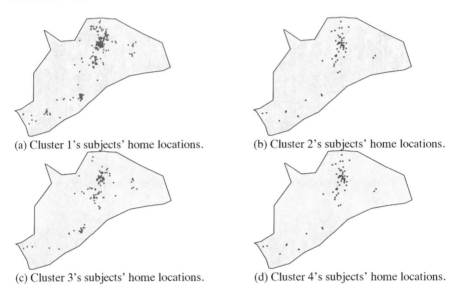

(a) Cluster 1's subjects' home locations.   (b) Cluster 2's subjects' home locations.

(c) Cluster 3's subjects' home locations.   (d) Cluster 4's subjects' home locations.

Fig. 8.24 The home locations of individuals in Fukui's various clusters. (**a**) Cluster 1's subjects' home locations. (**b**) Cluster 2's subjects' home locations. (**c**) Cluster 3's subjects' home locations. (**d**) Cluster 4's subjects' home locations

pattern to the destination (for repeat visitors) (for the repeat visitors). When a destination becomes well-known, there are usually a few common ways to get there.

8.5   Application

Using our framework, a variety of tourism applications can be developed. Here we demonstrate an application that can be valuable for urban planners, transportation managers, and tourism authorities. We developed a system interface that allows users to view and evaluate people's mobility patterns, particularly their travel behavior. We'd like to demonstrate it at two different levels: individual and aggregate. A user can monitor touristic, non-touristic, and commuting trajectories at the individual level, all of which allow the user to conduct further investigations into travel behavior or make more informed judgments in urban planning and transportation. Fig. 8.26 illustrates a screenshot of this user interface. The trajectories of four subjects in the Tokyo area, as well as the types of excursions and home and workplace locations, are displayed as an example.

A user can observe and evaluate tourist flows across cities or prefectures at the aggregate level. The user interface allows the user to study trends and flow distributions by observing the intake and outflow of any prefecture of interest for any set time of observation. Fig. 8.27 depicts a screenshot of our system interface at the aggregate level. The outflow of Tokyo is depicted in this image.

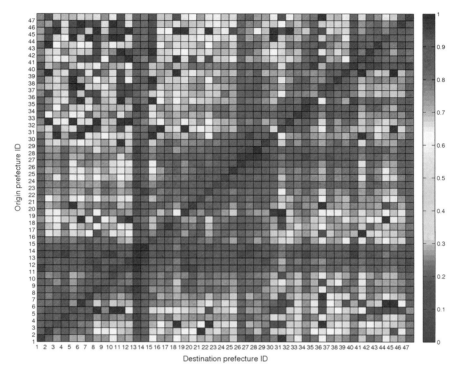

Fig. 8.25 O–D similarity matrix

8.6   Conclusion

Personal mobile phones' ubiquity and sensing capabilities can turn them into human probes—individual sensors that can monitor and record individual behavioral data, which can help reveal interesting human behavior phenomena that govern how cities function and how different urban elements operate collectively. In this chapter, we recognized and exploited this opportunistic sensing mechanism by conducting a study of tourist behavior utilizing extensive mobile phone GPS location records collected over the course of a calendar year from 130,861 users in Japan. We described a framework for detecting tourists using GPS traces, which enabled us to carry out our analysis. We particularly examined the number of trips made, the amount of time spent at destinations, the modes of transportation employed, and the relationship between personal mobility and tourist travel behavior. Some of the key findings the case study discussed in this chapter include the discovery of top origins and destinations of tourists, top flows between prefectures as well as within their own prefectures, top destination types for lengthier visits, most popular types of tourist destinations, popular transport mode behaviors, interesting link between personal mobility and travel behavior, and similarity observed between spatial

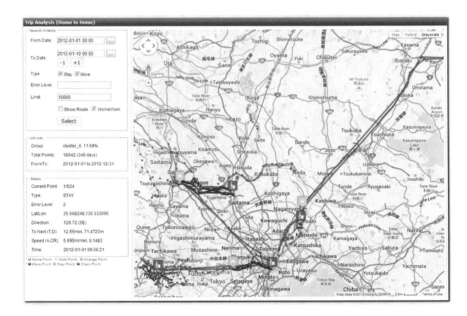

Fig. 8.26 A screenshot of our system interface, which allows the user to view various types of trips. The trajectories of four subjects are depicted in this example, along with the types of trips taken and the locations of their homes and workplaces. Commuting travels are shown in blue, touristic journeys are shown in red, and non-touristic trips are shown in green

Fig. 8.27 A screenshot of our user interface, which allows users to track tourist flows. This illustration depicts Tokyo's outflow

clusters of tourists. Moreover, we described an application that allows users to track people's mobility, which may be classified as commuting, non-touristic, or touristic journeys, as well as their home and work locations. This application interface also allows its user to monitor and analyze the inflow and outflow of any prefecture of interest over a specified time period. Urban planners, transit managers, and tourism authorities can potentially benefit from the application for their strategic planning and management.

Nonetheless, there are a few limitations of this research. To begin with, it's possible that the inferred touristic trips did not cover all genuine touristic journeys. Second, other modes of transportation were not considered in the investigation, such as cycling and taking the bus. Finally, the scope of travel was limited to domestic travels; thus, foreign trips were excluded from the analysis.

In terms of the size and longitudinality of the data for analysis, the study described in this chapter offers another approach of analyzing tourist behavior, with an advantage over traditional surveys and questionnaire studies. The study's findings also provide fascinating insights into tourist travel behavior. Real-time sensing and ICT-driven processes for sustainable tourism are among the research topics suggested and worth pursuing in the future.

References

1. Phithakkitnukoon S, Horanont T, Di Lorenzo G, Shibasaki R, Ratti C. Activity-aware map: Identifying human daily activity pattern using mobile phone data. LNCS. 2010;6219
2. González MC, Hidalgo CA, Barabási AL. Understanding individual human mobility patterns. Nature. 2008;453:779–82.
3. Song C, Qu Z, Blumm N, Barabási AL. Limits of predictability in human mobility. Science (80-). 2010;327(5968):1018–21.
4. Song C, Koren T, Wang P, Barabási AL. Modelling the scaling properties of human mobility. Nat Phys. 2010;6:818–23.
5. Phithakkitnukoon S, Smoreda Z, Olivier P. Socio-geography of human mobility: a study using longitudinal mobile phone data. PLoS One. 2012;7(6):e39253.
6. Alegre J, Pou L. The length of stay in the demand for tourism. Tour Manag. 2006;27(6): 1343–55.
7. Gokovali U, Bahar O, Kozak M. Determinants of length of stay: a practical use of survival analysis. Tour Manag. 2007;28(3):736–46.
8. Wu L, Zhang J, Fujiwara A. Dynamic analysis of japanese tourists' three stage choices. Transp Res Rec J Transp Res Board. 2012;2322(1):91–101.
9. Zheng Y, Li Q, Chen Y, Xie X, Ma W-Y. Understanding mobility based on GPS data. In: Proceedings of the 10th international conference on Ubiquitous computing–UbiComp '08; 2008. p. 312–21.
10. Ester M, Kriegel HP, Sander J, Xu X. A density-based algorithm for discovering clusters in large spatial databases with noise. In: Proceedings of the Second International Conference on Knowledge Discovery and Data Mining (KDD'96); 1996. p. 226–31
11. Breiman L. Random Forests. Mach Learn. 2001;45(1):5–32.

12. Statistics Bureau. 2006 establishment and enterprise census. Ministry of Internal and Communications. 2006;

13. Japanese Ministry of Land. Tourism Resources.

14. Breiman L. Degrowing tourism: decroissance, sustainable consumption and steady-state tourism. Anatolia Int J Tour Hosp Res. 2001;20(1):46–61.

15. Witayangkurn A, Horanont T, Ono N, Sekimoto Y, Shibasaki R. Trip reconstruction and transportation mode extraction on low data rate GPS data from mobile phone. In: Proceedings of CUPUM 2013: 13th International Conference on Computers in Urban Planning and Urban Management–Planning Support Systems for Sustainable Urban Development; 2013. p. 1–19.

An Outlook for Future Mobile Network Data-Driven Urban Informatics

9

Abstract

This chapter provides an outlook for future directions of mobile network data-based urban informatics, particularly in travel behavior research. It discusses a dynamic characteristic of mobile network data that continues to change its properties and potential values with the technological advancement, which in turn poses new challenges and exciting research opportunities. This may include new paradigms for data collection as an alternative to the telecom provided data, which is rarely accessible due to data privacy regulations. There is still a need for rebalancing between the data's utility and privacy for which data uncertainty and privacy algorithms are discussed. Future directions of mobile network data-based urban informatics will concern data mining techniques that help discover patterns and trends in trajectory data, including group movement pattern mining, trajectory clustering, and sequential pattern mining. Mining such patterns benefits travel behavior research with many applications, such as traffic detection, social gathering recognition, regional travel behavioral signature extraction, uncovering life/daily patterns, and unusual event detection.

Keywords

Mobile network data characteristics · Data utility · Data privacy · Data uncertainty · Trajectory data mining · Group movement pattern mining · Trajectory clustering · Sequential pattern mining

9.1 Mobile Network Data Characteristics

The use of mobile network data has already shown its benefits in urban informatics, particularly understanding human mobility in the form of travel behavior research and transportation modeling. Properties and potential value of the mobile network

© The Author(s), under exclusive license to Springer Nature Singapore Pte
Ltd. 2023
S. Phithakkitnukoon, *Urban Informatics Using Mobile Network Data*,
https://doi.org/10.1007/978-981-19-6714-6_9

data will continue to change with the technological advancement, mobile service usage behavior, and research efforts that leverage its utilization.

Collection of mobile network data relies on the underpinned communication technology, which has been changing consistently. Among the first contributions that the mobile network data has made in travel behavior studies are the revolutionary findings by researchers in statistical physics and computer science, who unveiled some hidden statistical patterns in travel behavior based on complex network theory in 2008 [1]. Soon after, more related research findings based on the same dataset were later published in 2010 [2, 3]. After a few more datasets became available for research, many more interesting discoveries and method developments have contributed to the field. From 2005–2010, mobile telecommunication technology was 2G transmission, which mainly serve voice and SMS services. The industry standard today is 4G/LTE service, which can support voice, SMS, and broadband internet access. With a higher user connectivity due to the internet usage today, the mobile network data's properties have changed. More connectivity implies more locational data samples of individual users. A higher spatial resolution can also be achieved with an additional use of Wi-Fi connectivity data, which can be collected through the operator's offered broadband services. The Wi-Fi data can provide an indoor locational data (i.e., access point locations), which can be useful for in-building mobility analysis that complements the cellular network-based trajectories. These technology advancements continuingly change the mobile network data's properties, and thus pose some new challenges and open up many interesting research opportunities to explore, examine, and develop methods and theories that describe travel behavior as well as extend state of the art in urban informatics.

With today's 4G/LTE technology, mobile network user's behavior has changed. Mobile device has been used for online services such as web browsing, chatting, playing games, shopping, and navigation, much more than the voice service. In addition, an internet voice call service (Voice over Internet Protocol: VoIP) such as LINE, WhatsApp, and Skype is becoming more popular as the user can make a voice call simply over the internet, so it's gradually replacing the conventional voice service. As a result, the percentage of mobile device internet traffic worldwide from 2015 to 2021 has increased dramatically [4]. Today, mobile internet traffic accounts for almost 55 percent of total web traffic [5]. The figure becomes even larger in mobile-first markets such as Asia and Africa. Not only the mobile device penetration rate increases, but also mobile device interaction or screen time. A survey conducted by Reviews.org in 2021 [6] reveals that people check their mobile phones 344 times per day i.e., once every four minutes, on average. Another survey in 2016 by Christensen et al. [7] found a median of 3.7 minutes per hour of screen time over a 30-day period. User behavior has changed in how they interact and spend time of their mobile devices. Shifting from voice call to internet connection has led to a more detailed locational mobile network data. Moreover, mobile app usage and web serving logs are monitored and collected by the telecom operators for billing as well as service improvement purposes. This usage behavior shifting has changed the mobile phone data's properties and potential value significantly. As a result, it poses new challenges to development of efficient methods, algorithms, and frameworks for inferring travel behavior with such detailed locational data as well as utilizing the internet usage data to complement the locational data analysis or even to better understand user's profile such as persona for data monetization opportunities for the telecom operators e.g., service personalization, customer segmentation, and personal personalized marketing. Furthermore, processing such large amount of data may exceed the capacity of traditional analytics tools. Processing issues may be caused by extracting meaningful insights from a massive data set. As the traditional data architectures may not be able to handle a large volume of mobile network data since they cannot cope with different characteristics of massive data sets e.g., velocity, variety, and veracity. Such data characteristics thus require a development of big data analytics platforms and cloud-based and edge computing frameworks as well as distributed computing architectures [8, 9].

Research efforts are reflected in analytical frameworks and data science approaches developed and aimed at using mobile network data in travel behavior studies and various other purposes. As mobile network connectivity can be considered as a function of population and observed spatial and temporal dynamics in a considered area (e.g., TAZ). Exploiting such spatial and temporal characteristics of mobile users requires spatio-temporal analysis and modeling to capture travel behavior and evaluate trends of human mobility. Besides the locational data, mobile communication logs can also be analyzed to reveal social networking behavior such as social network structures [10] and social network dynamics [11]. Exploiting both travel and social behaviors leads to a better understanding of their complex interrelationship, which appear to be indicative of each other [12, 13]. These research efforts will continue to increase the potential value of the mobile network data in years to come. Clearly, there are still research opportunities around the mobility and sociality aspects as well as their relationship that need to be explored through the use of mobile network data with ever-changing characteristics.

The data used in the case studies discussed in this book were 2013 CDR data from Senegal (Chaps. 1 and 2), 2007 CDR data from Portugal (Chaps. 4–6), and 2012 GPS data from Japan (Chaps. 7 and 8). These data allowed us to analyze travel behavior in detail with respect to the four-step models and activity-based approach, as well as compare them with census surveys and travel demand models. It is therefore important to put these data in context with today's mobile communication technology and changing behavior in mobile device usage. Those case studies' findings were based on periods of less intense use of mobile network services as compared with today. Availability of a more detailed and better-quality locational data of today's mobile network data can produce a more robust results and models for daily travel behavior and activities.

9.2   Data Collection

With GPS capability, today's smartphones can be used for a travel survey. When its GPS feature is enabled, the location of the device can be more precisely determined compared to the CDR data. It can improve the quality of locational information with accurate spatial-temporal trajectories of individuals as seen in our analyses discussed in chapters 7 and 8. Unlike a stand-alone GPS device, smartphone is ubiquitous in our modern society. Its high penetration and frequent user interaction easily allow a smartphone to be used as a ubiquitous sensor for individual behavior such as mobility and sociality. Apart from the GPS feature, a typical smartphone also includes other sensors such as accelerometer and magnetometer, which can also be used to collect a richer context of the traveler such as engaged activities [14] and detailed indoor mobility [15]. There has been a number of attempts to conduct a mobile phone-assisted travel survey. For example, the Probe Person survey system that combined built-in GPS with Internet Web diaries was developed to carry out a pilot survey in Japan in 2014 [16]. Another attempt was a survey conducted in 2014 in Austria with an Android-based smartphone app that combined accelerometer data and GPS speed and location information [17]. Although only small samples were collected by those surveys, they've paved the way and motivated future development of large-scale mobile phone-based travel survey. Today, there are already publicly available mobile apps that can be used for a travel surveys such as *rMove: Travel Survey* [18], *Household Travel Survey* [19], and *Sensor Log* [20]. Nonetheless, attaining a large dataset in terms of its sample size and observation period can still be very challenging, which needs further research for feasible frameworks and approaches. An attainable data size may not be as large as a typical CDR dataset provided by a telecom operator. However, its higher spatial resolution can produce a more detailed mobility that may compensate for its sample size. Nevertheless, it depends on the focus of the analysis. For example, a regional travel demand modeling may prefer a CDR data-based analysis, while a personalized mobility as a service (MaaS) modeling on the other hand may benefit more from a GPS trajectory data.

One of primary challenges that restricts the accessibility to mobile network data or mobile phone-assisted travel survey data is location privacy. Revealing such locations to a third party can pose user privacy violation issues as stipulated by GDPR [21] and HIPAA [22]. Mobile network data that are currently accessible for research are either available privately through an agreement with a telecom operator (which is mostly the case) or publicly available such as Nokia Mobile Data Challenge dataset [23], Orange Telecom Data for Development Challenge (D4D) [24, 25]. As de Montjoye et al. [26] discovered that just four spatiotemporal points of an anonymous mobile user may reveal the user's identity with 95% probability in a CDR data, there are needs and opportunities for research and development of efficient privacy preservation algorithms for trajectory data such as mobile network data that balances its utility and privacy.

9.3   Data Uncertainty and Privacy

From the travel behavior research's point of view, the goal is to extract individual trajectories from the mobile network data, so that those trajectories can be used to understand travel behaviors such as commuting patterns, transport modes, route choices, and so on. However, the mobile network data only provides a sample of the mobile user's actual movement. The location information is known only when the mobile device connects to the mobile network for services such as voice, SMS, and data (internet). The movement of a mobile user between two consecutive connected locations becomes unknown or so called *uncertain*. So, to obtain a trajectory with a complete movement of a mobile user as much as possible, we want to reduce the uncertainty of the data. On the other hand, to preserve a user's privacy, we aim to increase its uncertainty.

Like many other types of trajectories such as GPS coordinates of a moving object or social media check-in locations, a CDR-based trajectory is called uncertain trajectory. There are two scenarios in reducing uncertainty in a CDR trajectory, which formulate two different settings for our investigation, i.e., road-network constrained and free space movements. In a road-network constrained movement scenario, we assume that mobile users can only travel within a road network. So, the problem boils down to inferring the most likely route(s) that a mobile user could travel between a few sample points based on a set of uncertain trajectories. Basically, we want to fill in the missing sub-trajectories with coordinates along the most likely route(s) of a road network. One approach is to leverage the data from many other trajectories [27]. With the idea that other mobile users' trajectories may share the route(s) either exactly or partially, incorporating these trajectories can often supplement each other by making each of them more complete. In other words, its's possible to fill in an uncertain trajectory by cross-referring other trajectories on (or partially on) the same route(s), as shown in Fig. 9.1a where each shape represents each trajectory from the origin A to destination B. By incorporating other

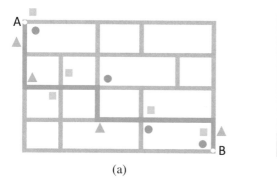
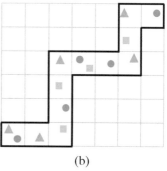

(a)   (b)

Fig. 9.1 Approaches for reducing uncertainty in CDR trajectories; (**a**) leveraging data from other trajectories in the road network constrained scenario, (**b**) partitioning a geospacer into uniform grids and then mapping trajectories onto these grids in the free space scenario

trajectories, the most likely route taken can be identified (highlighted in blue). By the same token, their own historical trajectories can also be used to infer the most likely routes as the most frequent and repeated paths, especially in the case where an observation period is sufficiently long. Another approach is to make use of the locations of used cellular towers along the possible routes [28]. More weight may be assigned to a more frequently connected cellular tower to guide our route inference model. Finding an optimal weighting scheme is still in open research question. Estimation of the likelihood of potential routes used by the mobile user still needs further investigation. One of the possible approaches is to interpolate the uncertain trajectory with coordinates along candidate routes, so that a likelihood of each candidate can be calculated as a distance from the interpolated points along the route to a set of used cell towers. With this approach, an optimal interpolation method needs to be developed and it is another interesting research topic to be explored.

In a free space movement scenario, a user's mobility is not constrained by a road network i.e., a travel does not follow paths in a road network, such as hiking a mountain, walking through a park, and so on. One approach is to partition a geospacer into uniform grids and then maps trajectories onto these grids, where the size of a grid may depend on the required inference accuracy [29]. A routable graph can then be constructed by connecting grids through which the trajectories pass based on some rules by considering starting/ending points and travel times. This yields a set of connected regions (grids) that form a directed routable graph where a node is a grid and direction and travel time between adjacent grids are inferred based on the trajectories passing the grids, as shown in Fig. 9.1b where each shape represents each user's uncertain trajectory.

Other approaches include pdf-based models where missing locations of a mobile user are described by a two-dimensional probability density function (pdf) and shape-based models where possible locations are bounded by geometric shapes (e.g., circle, lens, and polygon), which may be associated with some probability values [30]. There has been no claim about spatial pdf within a shape, which is worth an exploration. Another topic that is worth future investigation is how to identify precise locations of the mobile user from the used cell tower locations where there is no road network guideline in a case of CDR-based trajectories. Additional information may be needed such as hiking trails, walkable areas, transport mode used, and so on.

As opposed to reducing the uncertainty of a trajectory, there are techniques and approaches aim to preserve the privacy of a mobile user caused by the disclosure of the user's trajectories [31]. These techniques want to blur a user's location while ensuring the utility of the trajectory data. Two well-studied trajectory anonymization approaches are clustering-based [32] and generalization-based [33]. The clustering-based approach groups together k co-localized trajectories within the same time period to form a k-anonymized aggregate trajectory by utilizing the uncertainty of trajectory data. A given trajectory with an uncertainty threshold d is modeled by a horizontal disk with radius centered around the trajectory. The union of all such disks constitute the trajectory volume, as shown in Fig. 9.2a. Two trajectories are

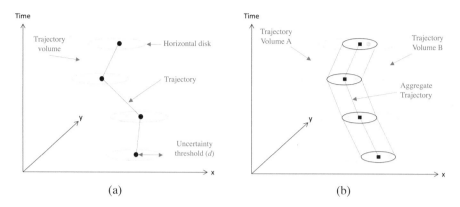

Fig. 9.2 Clustering-based approach; (**a**) trajectory uncertainty model, (**b**) an example of 2-anoymized co-localized trajectories

said to be co-localized with respect to the uncertainty threshold d, if the Euclidean distance between each pair of points in time is less than or equal to d. A cluster of k co-localized trajectories forms an aggregate trajectory where each locational points is an arithmetic mean of all location samples at that time. An example of 2-anonymized co-localized trajectories is shown in Fig. 9.2b where two trajectory volumes (A and B) represented in grey and dotted lines together form an aggregate trajectory represented by a sequence of black square markers whose trajectory volume represented with black lines is a bounding trajectory volume for trajectories A and B.

On the other hand, the generalization-based approach adopts the notion of k-anonymity [34] for anonymization of trajectories by generalizing a trajectory data set into a set of k-anonymized regions, and then for each k-anonymized trajectory uniformly selecting k points each from each anonymized region and linking a unique point from each region to reconstruct k trajectories. Conceptually, k-anonymity is a privacy standard to protect against identification of individuals by ensuring that every trajectory is indistinguishable from $k - 1$ other trajectories. So, a trajectory data is k-anonymous if each data record over quasi-identifiers appears at least k times. The k-anonymity property ensures that a given set of quasi-identifiers can only be mapped to at least k entities in the dataset. There are two main steps in the generalization-based approach: anonymization and reconstruction steps. The anonymization step is an iterative process that for each iteration an empty anonymity group G is created by randomly taking one trajectory sample from the data set. This anonymization process continues until G contains k trajectories. This step finishes when there are less than k remaining trajectories in the data set. An example of the anonymization step is shown in Fig. 9.3 where three trajectories Tr_1, Tr_2, and Tr_3 are generalized into a 3-anonymized trajectory. Trajectory Tr_2 is first added into an empty group G. Then, Tr_1 is added to G and the timestamped location records of Tr_1 and Tr_2 are generalized into a sequence of anonymized regions represented by shaded rectangles, as shown in Fig. 9.3b, which then forms an anonymization of

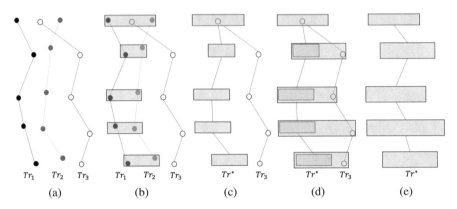

Fig. 9.3 Generalization-based approach: anonymization step; (**a**) original trajectories, (**b**) anonymity group $G = \{Tr_1, Tr_2\}$, (**c**) anonymization of G, (**d**) trajectory Tr_3 is added into G, (**e**) anonymization of G

Tr_1 and Tr_2, i.e., Tr^* as shown in Fig. 9.3c. Trajectory Tr_3 is also added, which then forms a sequence of anonymized regions for G as shown in Fig. 9.3d. The anonymization step finishes as soon as G contains three trajectories ($k = 3$). As the result, Fig. 9.3d shows the anonymization of G.

In the reconstruction step, from a given k-anonymized trajectory, k location records are uniformly selected in each of its anonymized region, as show in Fig. 9.4a. By linking a location record in each anonymized region, k trajectories are reconstructed, as shown in Fig. 9.4b. So, a reconstructed trajectory data set can then then released to the public or third parties for data analysis.

There will always be a need to rebalancing between trajectory data's utility and privacy to meet ever-changing demand in big data era. Future directions of data uncertainty and privacy inevitably concern development of efficient mechanisms for reducing data uncertainty as well as preserving privacy for location-based services, which will become more and more personalized, requiring more user semantics e.g., user preference and background information, rather than just some simple query parameters such as distance range. Dynamic in mobile network data characteristics also requires new techniques for data uncertainty reduction and privacy preservation that support more complex spatiotemporal queries beyond trajectories, such as persona, wealth index, living index, and so on.

9.4 Travel Behavior Pattern Mining

With the traditional four-step and activity-based models being employed as the main travel modeling approaches, there are still needs for efficient inference models and approaches for each step of the four steps i.e., trip generation, trip distribution, mode choice, and route assignment, as well as activity engaged by the mobile users, as already mentioned in the previous chapters. Future directions of mobile network

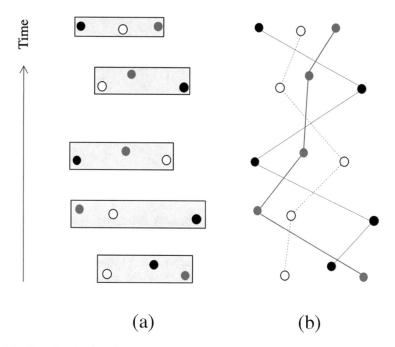

Fig. 9.4 Generalization-based approach: reconstruction step; (**a**) selection of timestamped location records, (**b**) reconstructed trajectories

data-based travel behavior research will concern data mining techniques that help recognize patterns and trends reflecting behavioral insights of how people travel. These techniques can be largely grouped into three categories, including group movement pattern mining, trajectory clustering, and sequential pattern mining.

9.4.1 Group Movement Pattern Mining

In order to analyze group travel behaviors of mobile users, a branch of research focuses on discovering group movement patterns aiming at identifying groups of mobile users that move together for a certain period of time, which can benefit various applications such as traffic detection, social gathering recognition, unusual event detection, and location-based services e.g., car-pooling. These group movement patterns include *flock* [35], *convoy* [36], *swarm* [37], and *gathering* [38].

A *flock* is a group of objects that move together within a circle of some user-specified size for at least k consecutive timestamps. The concept of k-co-movement patterns [39] was later introduced to improve the flock pattern's distance parameter limitation by using a top-down algorithm with free distance parameter. Nonetheless, the major concern with a flock is the predefined circular shape, which may not properly describe the shape of a moving group in reality. Its circular shape issue results in the so-called lossy-flock problem.

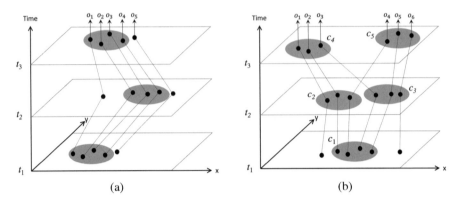

Fig. 9.5 Examples of group movement patterns: (**a**) flock, convoy, and swarm, (**b**) gathering

To address the issues of size and shape of a moving group, the *convoy* pattern is proposed to describe arbitrary shape of groups by employing density-based cluster-ing. A convoy requires a group of moving objects to be density-connected during k consecutive timestamps instead of using a circular shape.

While both flock and convoy have a restriction on consecutive time periods, a *swarm* is proposed as a general type of trajectory pattern that is a cluster of moving objects staying together for at least k timestamps, which can be nonconsecutive. Due to the fact that members of a group may not always be gathered, this swarm pattern thus relaxes the restrictions and allows the members to disappear for some time as long as they say together for at least k time periods.

In a case where moving objects join and leave an event (or cluster) temporarily, such as celebrations, parades, and so on, the *gathering* pattern is proposed to detect such group movement by further loosening the spatiotemporal constraints of the aforementioned patterns by allowing a group membership to evolve gradually. Formally, a cluster of a gathering must contain at least m participants, which refer to the moving objects appearing in at least k clusters of this gathering. Consequently, a gathering pattern requires geometric properties i.e., shape and location of the group to be stable as it is used to detect events. Hence, there is still a need for more robust and efficient modeling of group movement patterns that address the aforementioned constraints.

An example of flock, convoy, and swarm patterns with the required timestamps $k = 2$ is shown in Fig. 9.5a. A group of the objects $\{o_2, o_3, o_4\}$ forms a flock pattern from timestamps t_1 to t_3. Due to the fixed size of circular shape by the flock definition, the object o_5 is not included although it is a companion of the group. A convoy pattern, on the other hand, includes o_5 in the group because these objects $\{o_2, o_3, o_4, o_5\}$ are density-based connected throughout timestamps t_1 to t_3. Furthermore, a swarm pattern is formed by all five moving objects $\{o_1, o_2, o_3, o_4, o_5\}$ over the nonconsecutive timestamps t_1 to t_3. With $k = 2$ and $m = 3$, Fig. 9.5b shows an example of a gathering pattern that includes a group of clusters $\{c_1, c_2, c_4\}$ as it contains three participants all the time, while $\{c_1, c_3, c_5\}$ is not a gathering because c_5 is too far away from c_2 and c_3.

9.4.2   Trajectory Clustering

Trajectory clustering aims at grouping similar trajectories into clusters, so that we can identify crowd behavior, representative paths, common trends, or abnormal behaviors shared by different travelers. Clustering is done based on similarity or distance between trajectories, which are generally represented by feature vectors. To measure a similarity, clustering models require that the trajectories have a unified length. However, in most cases, trajectories have different lengths. Representing trajectories in a unified length with little loss of information is challenging as different trajectories contain different and complex properties, such as length, shape, sampling rate, and so on.

9.4.2.1 Trajectory Data Preparation

Therefore, there are approaches that try to represent original trajectory data in other space with a same length. For instance, a trajectory can be resented as a combination of basis trajectory using linear transformation algorithm [40]. Curve fitting is another approach to approximate trajectories by a parameterized quadratic curve, which is then used to distinguish different represented curves with the use of the direction of the last trajectory point as an additional information [41]. A trajectory data can also be approximated by a uniform cubic B-spline curve, so that we can obtain a representation that is capable of encoding both shape and spatiotemporal profile of the trajectory [42]. To distinguish the trajectories with similar shapes, the lengths of trajectories are considered. Additionally, vector fields can also be used to induce a notion of similarity between trajectories with the idea that movement trends in the trajectories are modeled as flows within multiple vector fields and the vector filed itself differentiates the clusters [43]. A statistical procedure for identifying a smaller set of linearly uncorrelated variables called principal components by orthogonal transformation from a larger set of data like Principal Component Analysis (PCA) can also be used to represent a trajectory with its computed coefficients [44]. As a trajectory can be considered as a time series that can be represented in a frequency domain, therefore a trajectory can also be modeled as a fixed length vector comprised of Fourier coefficients by using the Discrete Fourier Transformation (DFT) [45, 46]. Nonetheless, representing trajectories in a unified length causes a loss of information. So, there is still a pressing need for more efficient methods and approaches worth mentioning here as one of the future research directions.

9.4.2.2 Distance Measurements

To measure similarities among different trajectories with a unified length, Euclidean distance can be used, where trajectories are represented as a vector [47], as follows.

$$D_{Euc}(A, B) = \frac{1}{N} \sum_{n=1}^{N} \sqrt{\left(a_n^{lat} - b_n^{lat}\right)^2 + \left(a_n^{lon} - b_n^{lon}\right)^2}, \quad (9.1)$$

where (a_n^{lat}, a_n^{lon}) and (b_n^{lat}, b_n^{lon}) are the nth points (geographic coordinates) of trajectories A and B, respectively, and N is the total number of coordinates. The computational complexity of the Euclidean distance is $O(n)$. Trajectories can also be treated as samples of a probabilistic distribution for which Bhattacharyya distance is used to measure how closely of two probability distributions based on quantized directions of the trajectories' coordinates [48], as follows.

$$D_{Bha}(A, B) = -\ln\left(\sum_{t=1}^{T} \sqrt{a_t \cdot b_t}\right), \quad (9.2)$$

where a_t and b_t are the quantized directions that are used to measure the separability of trajectories A and B, respectively. The Bhattacharyya distance's computational complexity is the same as Euclidean distance, which is $O(n)$.

There are also approaches to measure similarities among different trajectories without unifying the lengths. Hausdorff distance, for instance, measures the similarities by considering how close every coordinate of one trajectory to some coordinates of the other one [49, 50], as follows.

$$D_{Hau}(A, B) = max\{d(A, B), d(B, A)\}, \quad (9.3)$$

where

$$d(A, B) = \max_{a \in A} \min_{b \in B} \|f_a - f_b\|, \quad (9.4)$$

$$d(B, A) = \max_{b \in B} \min_{a \in A} \|f_b - f_a\|, \quad (9.5)$$

and $\|f_a - f_b\|$ is a Euclidean distance between flow vectors that are used to represent trajectories A and B. The computational complexity of Hausdorff distance is $O(mn)$.

Another approach is to approximate trajectories A and B with metric curves i.e., continuous map from unit interval into a metric space S, and then use Fréchet distance to measure a similarity between the two curves by taking into account the location and time ordering [51]. Fréchet distance is defined as follows.

$$D_{Fre}(A, B) = \inf_{\alpha,\beta} \max_{t \in [0, 1]} \{d(A(\alpha(t)), B(\beta(t)))\}, \quad (9.6)$$

where t is the measure of time, d is a distance function of S, while α and β are non-decreasing reparameterization mapping functions of the trajectories A and B, respectively. The Fréchet distance's computational complexity is $O(mn)$.

By not considering the lengths and time ordering of the trajectories, Dynamic Time Warping (DTW) distance measures similarity by matching the vertices (coordinates) from one trajectory to vertices on the other such that the summed distance between matched vertices is minimized. DTW distance basically searches the optimal warping (i.e., minimum distance) path between two trajectories specified by using Euclidean or other distance function such as Harversine formula that is typically used for calculating a geographic distance between points on earth based on the spherical law. DTW distance is defined as follows.

$$D_{DTW}(A, B) = \min_{f} \frac{1}{n} \sum_{i=1}^{n} d\left(a_i, b_{f(i)}\right), \qquad (9.7)$$

where d is a distance function, trajectories A and B have n and m points, respectively, and all mappings $f : [1, n] \to [1, m]$ satisfy the conditions that $f(1) = 1$, $f(n) = m$, and $f(i) \leq f(j)$, for all $1 \leq i \leq j \leq n$. Nonetheless, since DTW only considers vertices of the trajectory and ignores the edges between these vertices, the DTW is sensitive to the relative sampling rates of the trajectories. The computational complexity of DTW distance is $O(mn)$.

Another distance metric that does not require unified lengths of trajectories is the Longest Common Subsequence (LCSS) that aims at finding the longest common subsequence in all sequences, and the length of the longest subsequence can be considered as the similarity between two given trajectories with different lengths. LCSS distance is defined as follows.

$$LCSS(A, B) = \begin{cases} 0, \text{ if } A \text{ or } B \text{ is empty} \\ 1 + LCSS(Head(A), \ Head(B)) \text{ if } D_{Euc}(a_N, \ b_M) < \varepsilon \text{ and } |N - M| < \delta \\ \max\left(LCSS(Head(A), \ B), \ LCSS(A, \ Head(B))\right), \text{ otherwise} \end{cases} \qquad (9.8)$$

where $Head(A)$ indicates the first $N{-}1$ points of the trajectory A and $Head(B)$ denotes the first $M{-}1$ points of B, and finally, $D_{LCSS}(A, B) = 1 - \frac{LCSS(A, B)}{\max(N, M)}$. So, the LCSS distance has its computational complexity of $O(mn)$.

Distance metrices are used for similarity measurement, which is an integral part of the trajectory clustering. It is therefore critical to choose an optimal distance according to the application. For instance, Euclidean and Bhattacharyya distances are suitable for trajectories with unified lengths, while methods like Hausdorff, Fréchet, DTW, and LCSS distances can be used without concerning trajectory lengths. Nonetheless, the shortcomings of the aforementioned approaches leave the door open for future development of a more robust method that is efficient in both performance and computational complexity.

9.4.2.3 Clustering Models

As most mobile network data-based trajectory datasets are unlabeled data i.e., no ground truth available, unsupervised algorithms are employed. Clustering is a method that draws a hidden structure that describes internal relationships between unlabeled trajectories, which can be grouped into three main categories based on methodological characteristics, including densely clustering, hierarchical clustering, and spectral clustering.

9.4.2.3.1 Densely Clustering

Densely clustering is a procedure that groups together the closely points based on their centroids. Density-based spatial clustering of applications with noise (DBSCAN) [52] is a clustering method that is inspired by this idea and it has been widely applied in trajectory clustering. Given a point p and distance threshold ε, any points inside a circular area centered at p with a radius ε are called *directly reachable* to p. Furthermore, points $\{q_1, q_2, q_3, \ldots, q_n\}$ are reachable to p if there is a path that q_1 is directly reachable to p and each q_{i+1} is directly reachable to q_i, hence it leaves other points as the outliers, as shown in Fig. 9.6a. When it comes to dealing with trajectory data, applying DBSCAN to clustering trajectories as a whole may not detect similar portions of the trajectories [53]. A trajectory may have a long and complicated path. Even though some portions of trajectories show a common behavior, the whole trajectories may not. So, trajectories are partitioned and substituted by sub-trajectories, which are then clustered. These clustered sub-trajectories are grouped together at the last step. An example resulting clustered trajectories is shown in Fig. 9.4b. Due to the data characteristics of trajectories, the distance can be measured by a combination of perpendicular distance, parallel distance, and angle distance with some weight [54]. The clusters are then used to form the core trajectories, which later serve as a guideline for classifying new coming trajectory. For instance, an average of all trajectories' coordinates of a cluster is considered as a new point, while the clusters are represented by all averaged points.

In addition to DBSCAN, other well-studied algorithms that belong to densely clustering paradigm include k-means, Expectation-Maximization (EM), and fuzzy c-means (FCM). As k-means algorithm seeks to minimize the average squared distance between points in the same cluster, the k-means approach clusters trajectories by searching centroids of clusters, repeatedly until converged i.e., reaching a local minimum [43]. Its well-known drawback is the assumption that

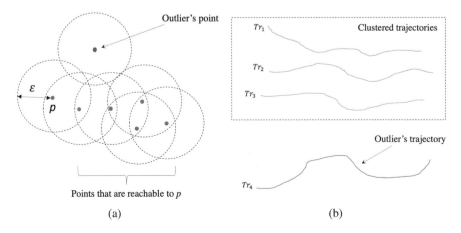

Fig. 9.6 Illustrations of the DBSCAN algorithm's concept for clustering (**a**) data points and (**b**) trajectories

the number of clusters (k) and initial clusters' centroids are given. So, the EM algorithm can be employed to improve the performance of k-means by solving optimization problem iteratively [55]. However, EM algorithm may suffer from its slow convergence. FCM algorithm are similar to k-means except for the "hardness" level of clustering. Unlike the k-means where the data points exclusively belong to one cluster, the FCM algorithm allows the data points to be assigned to multiple clusters with a likelihood. FCM algorithm employs parameters to measure the level of cluster fuzziness for each trajectory, and iteratively moves the cluster centroids to the "correct" locations by minimizing a criterion function measuring the quality of a fuzzy partition [56]. Nonetheless, the drawbacks of the FCM algorithm include sensitivity to initial selection of the centroids, slow convergence, and tendency to become stuck in a local optimum [57].

There is still room for improvement regarding the densely clustering methods. Future investigation may look into how to improve the models by incorporating other contextual information in addition to the spatial attributes, such as time of day, day of week, social context, or demographic information. Another direction is to explore ways to overcome their difficulty to handle the overlapping clusters (e.g., trajectory crossover) the loss of local information of trajectories.

9.4.2.3.2 Hierarchical Clustering

Hierarchical clustering is a method that produces hierarchal clusters of trajectories by constructing and analyzing a dendrogram—i.e., a tree-like structure that explains the relationship between all the trajectories (each treated as a data point) in the system. Nested clusters are generated by starting with a single data point as a cluster to a single cluster having all the data points or vice versa, where merges and splits are based on cluster distance. The clustering outcome is obtained by considering linkage strategy and similarity (or distance) measure.

A dendrogram is constructed iteratively by either merging (agglomeration) or splitting (division) of data points. So, there are two types of dendrogram construction: agglomerative and divisive clustering [58]. The agglomerative clustering is a "bottom-up" approach that starts with each data point as a single cluster. Distance between all clusters is calculated and used for merging two clusters with a minimum distance between them i.e., most similar clusters. A distance matrix may be used in this stage to keep track on cluster distance, so it is updated correspondingly as the process continues (moving up the hierarchy) until only one cluster remains or a requisite number of clusters is achieved. Linkage criteria are used as ways to determine proximity between clusters (or cluster distance). Different linkage criteria used lead to different hierarchical clustering models. In some cases, proximity can be determined as the distance between the two furthest points of the clusters by which the model is called *complete linkage* method. On the way, if the proximity is defined as the distance between the two nearest points of the clusters, it is called *single-linkage* method. Furthermore, the proximity can be an average distance between all the pairs of data points between the clusters. The average distance approach can further deviate into two methods based on weighting scheme namely unweighted average linkage clustering (or Unweighted Pair Group Method with

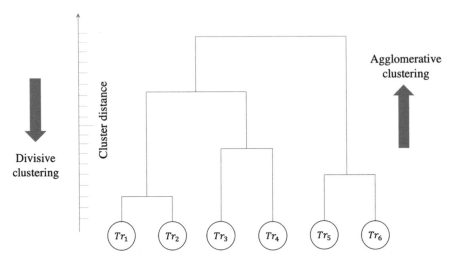

Fig. 9.7 An example of a dendrogram for hierarchical clustering of trajectories (Tr_1, Tr_2, ...). Dendrogram can be constructed based on cluster distance by two approaches: agglomerative clustering (bottom-up approach) and divisive clustering (top-down approach)

Arithmetic Mean: UPGMA) and weighted average linkage clustering (or Weighted Pair Group Method with Arithmetic Mean: WPGMA) [59].

Divisive clustering is a "top-down" approach, which is just an opposite of the agglomerative clustering, where its process starts with all data points in one single cluster and splits are performed recursively as we move down the hierarchy until a requisite number of clusters is achieved as we move down the hierarchy to the point where all data points exist as single clusters. This approach starts by splitting of the largest cluster into two clusters based on some cluster distance. The process continues until every data point is separated. A distance matrix is generally used in this process to keep track on cluster distance measures—e.g., an $n \times n$ matrix that contains cluster distance of n trajectories from all the other trajectories including itself is calculated, and then two most distant trajectories are chosen as the centers of new clusters, and further process is carried out. Divisive clustering is more conceptually complex than that of agglomerative clustering as it requires another flat clustering method (i.e., splitting a cluster into two clusters) as a subroutine.

An example of hierarchical clustering models is shown in Fig. 9.7 that illustrates both agglomerative and divisive clustering approaches, where merges and splits are based on cluster distance. In addition to spatial attributes, temporal attributes such as temporal proximity i.e., whether the trajectories occurred within a certain time frame from each other, have also been used to define the trajectory clusters in hierarchical clustering [60]. Moreover, hierarchical clustering approaches can also be improved by considering spatio-temporal information such as direction, speed, and time [61]. By adopting the concept of agglomerative hierarchical clustering, we can also cluster individual user's stay points instead of trajectories where links are generated between any two clusters in a chronological direction according to the time serial of the stay points, which altogether form individual travel sequences with significant locations based on visit frequency that initiatively reflect on personal interest in different types of places e.g., shopping malls, restaurants, parks, and so on [62].

One of the most attracting features of hierarchical clustering is that the number of clusters does not need to be preassigned, which makes it a flexible approach. In addition, the dendrogram helps understand the big picture as well as the groups in our trajectory data. However, the main disadvantage is that once a merge or split process is performed, it is difficult to make adjustments in a cluster. So, selection of the processes i.e., linkage strategy, distance metric, and so on, is crucial and could lead to low quality clusters. Other limitations of the hierarchical clustering methods are sensitivity to noise and outliers and difficulty when handing with different sizes of clusters. Identifying the optimal number of clusters for hierarchical clustering is still an open research question. Future research may investigate these limitations in the context of trajectory clustering for uncovering hidden travel behavior, as well as try to address the common drawbacks of the hierarchical clustering approaches that are hierarchical reference spots and consideration of sequence.

9.4.2.3.3 Spectral Clustering

Spectral clustering is a method derived from the graph theory and aims at identifying communities of nodes in a graph based on the edges connecting them. In the context of trajectory clustering, the trajectories and distance between them can be represented by nodes and edges, respectively. The graph is cut into sub-graphs to classify trajectories, where each sub-graph represents its own cluster. Spectral clustering approach makes use of the *spectrum* of the matrix, which refers to the set of its eigenvalues of the distance matrix of the trajectory data to perform dimensionality reduction before clustering in fewer dimensions.

The three major steps involved in spectral clustering are constructing a distance matrix, projecting trajectory data onto a lower-dimensional space, and clustering the data. It starts with forming a distance matrix for a given a set of trajectories. The distance matrix is then transformed into an affinity matrix A, which is a matrix of mutual similarities between the trajectories, and it can be constructed as follows.

$$A_{ij} = e^{\left(\frac{-d_{ij}}{2\sigma^2}\right)}, \tag{9.9}$$

where d_{ij} is the distance between trajectories i and j. With the affinity matrix, a Laplacian matrix L is computed as follows.

$$L = D - A, \tag{9.10}$$

where D is the diagonal matrix, i.e., $D_{ij} = \sum_j A_{ij}$. Eigenvalues and eigenvectors of L are then computed from which the k eigenvectors with the largest eigenvalues are retained. Clustering is finally performed on the data points in k-dimensional space.

One of the main advantages of the spectral clustering is the fact that it makes no assumption on the forms of clusters. As opposed to clustering techniques, like *k*-means that generally assumes that the clusters are spherical or round. In spectral clustering, the clusters do not follow a fixed shape or patterns. This suggests that the algorithm could be effective for data of different shapes and sizes, like trajectory data. When compared with other clustering algorithms, the spectral clustering is computationally effective for spare datasets [63]. However, it is computationally expensive for large datasets due to its computational cost of computing eigenvalues and eigenvectors and clustering that is performed on these vectors.

In the context of trajectory clustering, the spectral clustering approach has been employed and modified to improve its performance by feature learning and graph construction [64]. The concept of hybrid linear modeling has also been adopted to improve the performance of spectral clustering of trajectories by approximating a trajectory dataset with a mixture affine subspaces [65]. As a result, the approach only requires linear storage and takes linear running time in the size of the data. Representing trajectories with feature covariance matrix or *covariance descriptor* [66] in a spectral clustering can allow the calculation of a distance between trajectories of different lengths. Covariance descriptor basically enables us to determine the distance between two trajectories by representing the trajectories by their features and their covariance matrix of the feature matrices [67]. Hence, it helps avoid considering different lengths of trajectory data.

By comparison, densely clustering models group trajectories mostly by distance metrics which result in a clustering of trajectories solely by spatial information, while hierarchical clustering models addresses this limitation by considering more attributes in each hierarchy level. Nonetheless, this operation of the hierarchical clustering approach is computationally expensive. To address this, spectral clustering approach computes internal relationships by analyzing the affinity matrix, while reduces computational cost by processing all trajectory data together. Spectral clustering has its own limitation that its model can be well defined only for the non-negative affinities between trajectories [68]. Another limitation of the spectral clustering approach is the requirement of unified lengths of trajectories.

There are still many exciting areas to explore for future research in trajectory clustering, especially for mobile network-derived trajectory data. For instance, it is worth wide exploring other useful attributes of trajectories such as time of day, speed, and proximity to landmarks, and methods for extracting effective trajectory features with high discriminatory power. Moreover, future research may seek alternative approaches to identify a representative trajectory that best describes each cluster e.g., *k*-means identifies a representative center for each cluster. It is also worth wide exploring other forms of trajectory representation that allows a more accurate similarity measurement, such as graph-based representation where vertices and edges are utilized for similarity measurement [69].

9.4.3   Sequential Pattern Mining

Sequential pattern refers to a set of individual trajectories that share a common sequence of locations over a similar travel time interval. A sequential pattern is said to be detected when there is a common sequence of locations between trajectories over a similar time interval in which travel sequence's locations do not have to be consecutive. Therefore, two notions that are central for a sequential pattern detection are *location* and *travel time*. For a given set of trajectories A and B whose location sequences and travel times are as follows.

$$Tr_A : l_1 \xrightarrow{10\,min} l_2 \xrightarrow{15\,min} l_3 \xrightarrow{55\,min} l_4, Tr_B : l_1 \xrightarrow{22\,min} l_3 \xrightarrow{50\,min} l_4 \xrightarrow{45\,min} l_5.$$

It is said that Tr_A and Tr_B share a common sequence $l_1 \rightarrow l_3 \rightarrow l_4$, as the locational sequences and travel times are similar. This occurrence of common sequence is called a *support*. Sequential pattern mining is a process that extracts these sequential patterns whose support exceeds a predefined minimal support threshold. Suppose that l_1, l_3, and l_4 are central rail station, imperial temple, and museum, respectively. It may be interpreted as a typical behavior of tourists that visit a major attraction from the railway station and spend time there about an hour before visiting a nearby museum. Discovering such kind of patterns has various benefits, such as traffic management, travel recommendation, next-location prediction, daily pattern understanding, travel behavior similarity estimation, trajectory compression, and anomaly detection.

In order to detect a sequential pattern in trajectories, we first need to identify common locations, such as restaurant, monument, school, which is a place assigned with a unique ID—i.e., locations with the same ID are considered as common. In the case of CDR-based trajectories, cellular tower IDs can be used, so that a sequential pattern is represented as a sequence of the cellular tower IDs. However, for the GPS trajectories where each point is characterized by a pair of GPS coordinates, which rarely repeat their exact locations and hence it's unlikely to find common locations. To address this issue, algorithms and methods are thus needed to identify common locations or so called regions of interest (ROI), such as by adopting techniques in data clustering [69] and machine learning [70], or using points of interest (POIs) information [71], for instance.

For a GPS-based trajectory data, points from different trajectories first need to be classified into ROIs. An early method [72] was inspired by the line simplification technique or Douglas-Peucker algorithm [73], which identifies key points shaping each trajectory, resulting in simplified line segments. ROI is then represented by each simplified line segment from which the support is counted by grouping the fragments of a trajectory that are close to each ROI. Travel time is not considered in this approach. A more recent approach is to cluster points from different trajectories into a ROI, each represented by a cluster ID. As a result, a trajectory is transformed into a sequence of cluster IDs, which are comparable across different trajectories (similar to cellular tower IDs in the case of CDR-based trajectories), allowing us to mine sequential patterns from these sequences by using exiting sequential pattern mining algorithms, such as GSP (Generalized Sequential Pattern) [74], SPADE (Sequential Pattern Discovery using Equivalence Classes) [75], FreeSpan (Frequent Pattern-projected Sequential Pattern Mining) [76], PrefixSpan (Projected Sequential

Pattern Mining) [77], and CloSpan (Closed Sequential Pattern Mining) [78]. Here we briefly discuss these widely used algorithms to outline their differences.

Among the earlier approaches to sequential pattern mining is Apriori-like i.e., based on the Apriori property [79], which states that all non-empty subset of frequent itemset must be frequent. So, any *super-sequence* of a nonfrequent sequence cannot be frequent. Let sequence s be an ordered list of items, denoted by $s = \langle i_1, i_2, \ldots, i_k \rangle$. Given two sequences $\alpha = \langle a_1, a_2, \ldots, a_n \rangle$ and $\beta = \langle b_1, b_2, \ldots, b_m \rangle$, α is called a subsequence of β, denoted as $\alpha \subseteq \beta$, if there exit integers $1 \leq j_1 < j_2 < \ldots < j_n \leq m$ such that $a_1 \subseteq b_{j_1}, a_2 \subseteq b_{j_2}, \ldots, a_n \subseteq b_{j_n}$. We call β a super-sequence of α. For example, $\alpha = \langle l_1, l_3, l_4 \rangle$ and $\beta = \langle l_1, l_2, l_3, l_4, l_5 \rangle$. So, the Apriori property helps reduce the search space. Item can be considered as a ROI, so an itemset is a sequence of ROIs in the case of trajectory data. Based on this heuristic approach that utilizes prior knowledge of frequent itemset properties, the GSP algorithm adopts a multiple-pass, candidate-generation-and-test approach. The first scan (or pass) yields a set of all single item frequent sequences, which are then used as a *seed set* of sequential patterns for the next pass. This seed set is used as new potential patterns or *candidate sequences* in each pass, where the number of items (called *length*) in each candidate sequence increases by one for each pass. Each pass finds the support for each candidate sequence. All candidates whose support meet minimum support threshold form a set of the newly found sequential patterns, which then become the seed set for the next pass. The process continues until there is no new sequential pattern is found in a pass, or no candidate sequence can be generated.

SPADE is another algorithm that takes the Apriori-based approach with a vertical formatting method, as opposed to the GSP algorithm whose approach is considered as horizonal formatting that is not suitable for mining long sequential patterns due to its multiple scans of the database required by generation of a large set of candidate sequences. It improves on the GSP algorithm by using a support counting method based on the ID list structure which is more efficient method. As the result, the SPADE algorithm scales linearly with respect to the database size, number of sequences, and other database parameters.

Its stepwise approach of an Apriori-based algorithm causes expensive candidate sequence generation and test. To address this issue, a pattern-growth approach called FreeSpan algorithm uses frequent items to recursively project sequence databases into a set of smaller projected databases and grow subsequence fragments in each projected database. As the result, the data and the set of frequent patterns to be tested are portioned by this process, which confines each test to a corresponding smaller projected database.

Although the FreeSpan is proven to mine the complete set of patterns faster than the GSP algorithm, its projection has to keep the whole sequence in the original database because a subsequence may be generated by any combination in a sequence, which can be costly as the growth of a subsequence is explored at any split in a candidate sequence. Based on the pattern-growth approach used by the FreeSpan, the PrefixSpan algorithm deals with the aforementioned issue by examining only the prefix subsequences and projecting only their corresponding postfix subsequences into projected databases. So, the sequential patterns are grown by exploring only local frequent patterns in each projected database.

Another alternative pattern-growth-based algorithm is CloSpan, whose approach is to mine only frequent *closed subsequences* instead of the complete set of frequent subsequences. Closed subsequences refer to those containing no super-sequence with the same support. As the result, the CloSpan avoids the unnecessary traversing of search space with its backward sub-sequence and backward super-sequence methods, which consequently allow some sequences to be absorbed or merged and hence reduce the search space growth.

In a road network setting, trajectories can be first mapped into a road network using map-matching algorithms [80], then a sequential pattern mining can be performed by considering road segments IDs as items that form sequences to be mined. Sequential pattern mining algorithms have been applied to trajectory data that have generated interesting findings for travel behavior research, such as discoveries of sequential patterns of ROIs in a city [81] and significant places-based life patterns [82]. From the algorithmic point of view, there has been a number of methods developed to improve efficiency of sequential pattern mining for trajectory data, such as a graph-based sequence-matching algorithm for similarity estimation of different users' trajectories [83, 84], trajectory data compression [85], and suffix tree-based method for reducing the candidates of sub-trajectories when estimating the travel time of a trajectory [86].

Therefore, two main possible future research directions are (1) applying sequential pattern mining algorithms to trajectory data to discover insightful travel behavior patterns that extends the knowledge boundaries and (2) developing sequential pattern mining algorithms to improve the exiting methods for trajectory data, especially for mining long sequential patterns.

9.5 Conclusion

The collection and analysis of mobile network data represent dynamic areas of data and research that benefit urban informatics, particularly the travel behavior research in terms of transport planning and modeling. Properties and potential values of the mobile network data will continue to change with the technological advancement in telecommunication, mobile service usage behavior, and research efforts that leverage its utilization. This, in turn, poses new challenges and opens up countless research opportunities to explore, examine, and develop methods and theories that help provide a better understanding of travel behavior, as well as extend state of the art in urban informatics.

As one of the primary challenges that restricts the accessibility to mobile network data is data privacy regulation e.g., GDPR and HIPAA, the future data collection may need to explore and utilize the multisensory capabilities of today's smartphones. Attaining a large dataset can still be challenging, which needs further research for feasible frameworks. From the travel behavior study's point of view, one of the goals is to extract individual trajectories from the mobile network data, so that the extracted trajectories can be used for the analysis. However, the CDR data only provide the users' location information when the mobile device connects to a cellular network. The movement of the mobile user between two know locations becomes unknown or uncertain. This poses challenges and research opportunities to reduce this uncertainty in mobile network data. On the other hand, to preserve the user's privacy, the goal is to increase its uncertainty—especially, in the case of a mobile network GPS-based trajectory data whose mobility information is much more detailed than a CDR-based. Algorithms and techniques have been developed for dealing with mobile network data's uncertainty and privacy issues. Nonetheless, there will always be a need to rebalancing between data's utility and privacy to meet ever-changing demand, which requires robust algorithms.

In addition to the development of efficient inference models and approaches for each step of the four-step and activity-based models as discussed in the previous chapters, future directions of mobile network data-based travel behavior research will concern data mining techniques that help recognize patterns and trends reflecting behavioral insights of how people travel. These techniques can be largely grouped into three categories including group movement pattern mining, trajectory clustering, and sequential pattern mining.

Different group movement patterns and their constraints are discussed. Mining such group movement patterns can benefit traffic detection, social gathering recognition, unusual event detection, and location-based services. Trajectory clustering aims at grouping similar trajectories into clusters, so that we identify crowd behavior, representative paths, common trends, or abnormal behaviors shared by different travelers. Trajectory clustering requires data preparation, metrices for distance (similarity) measurement between trajectories, and clustering models. Based on methodological characteristics, clustering models are divided into main categories including densely clustering, hierarchical clustering, and spectral clustering. Discovering sequential patterns in trajectory data has various benefits, such as traffic management, travel recommendation, next-location prediction, daily pattern understanding, travel behavior similarity estimation, trajectory data compression, and anomaly detection. Existing sequential pattern mining algorithms' approaches and limitations are discussed, which provide an outlook for possible future research directions in urban informatics.

References

1. González MC, Hidalgo CA, Barabási AL. Understanding individual human mobility patterns. Nature. 2008;453:779–82. https://doi.org/10.1038/nature06958.
2. Song C, Koren T, Wang P, Barabási AL. Modelling the scaling properties of human mobility. Nat Phys. 2010;6:818–23. https://doi.org/10.1038/nphys1760.
3. Song C, Qu Z, Blumm N, Barabási AL. Limits of predictability in human mobility. Science (80-). 2010;327(5968):1018–21. https://doi.org/10.1126/science.1177170.
4. Clement J. Share of global mobile website traffic 2015–2021. Statista. 2022. https://www.statista.com/statistics/277125/share-of-website-traffic-coming-from-mobile-devices/. Accessed Mar. 03, 2022.

5. Ceci L. Mobile internet usage worldwide–statistics & facts. Statista. 2022. https://www.statista.com/topics/779/mobile-internet/#dossierKeyfigures. Accessed Mar. 03, 2022.
6. Wheelwright T. 2022 cell phone usage statistics: how obsessed are we? *Reviews.org*, 2022. Trevor Wheelwright. Accessed Mar. 03, 2022.
7. Christensen MA, et al. Direct measurements of smartphone screen-time: relationships with demographics and sleep. PLoS One. 2016;11(11):e0165331. https://doi.org/10.1371/journal.pone.0165331.
8. Casadei R, Fortino G, Pianini D, Russo W, Savaglio C, Viroli M. A development approach for collective opportunistic edge-of-things services. Inf Sci (Ny). 2019;498:154–69. https://doi.org/10.1016/j.ins.2019.05.058.
9. Liu J, Wan J, Zeng B, Wang Q, Song H, Qiu M. A scalable and quick-response software defined vehicular network assisted by Mobile edge computing. IEEE Commun Mag. 2017;55(7): 94–100. https://doi.org/10.1109/MCOM.2017.1601150.
10. Phithakkitnukoon S, Dantu R. Mobile social group sizes and scaling ratio. AI Soc. 2011;26(1): 71–85. https://doi.org/10.1007/s00146-009-0230-5.
11. Leskovec J. Dynamics of large networks. Carnegie Mellon University; 2008.
12. Phithakkitnukoon S, Smoreda Z, Olivier P. Socio-geography of human mobility: a study using longitudinal mobile phone data. PLoS One. 2012;7(6):e39253. https://doi.org/10.1371/journal.pone.0039253.
13. Phithakkitnukoon S, Smoreda Z. Influence of social relations on human mobility and sociality: a study of social ties in a cellular network. Soc Netw Anal Min. 2016;6:1. https://doi.org/10.1007/s13278-016-0351-z.
14. Phithakkitnukoon S, Dantu R. ContextAlert: context-aware alert mode for a mobile phone. Int J Pervasive Comput Commun. 2010;6(3):1–23. https://doi.org/10.1108/17427371011084266.
15. Gozick B, Subbu KP, Dantu R, Maeshiro T. Magnetic maps for indoor navigation. IEEE Trans Instrum Meas. 2011;60(12):3883–91. https://doi.org/10.1109/TIM.2011.2147690.
16. Asakura Y, Hato E, Maruyama T. Behavioural data collection using mobile phones. In: mobile technologies for activity-travel data collection and analysis; 2014. p. 17–35.
17. Nitsche P, Widhalm P, Breuss S, Brändle N, Maurer P. Supporting large-scale travel surveys with smartphones–a practical approach. Transp Res Part C Emerg Technol. 2014;43(2):212–21. https://doi.org/10.1016/j.trc.2013.11.005.
18. RSG. rMove: travel survey. Google play, 2022. https://play.google.com/store/apps/details?id=com.rsginc.rmove&hl=en_US&gl=US. Accessed Mar. 09, 2022.
19. ETC Institute. Household travel survey. Google play, 2022.
20. hfalan. Sensor log. Google play, 2022. https://play.google.com/store/apps/details?id=com.hfalan.activitylog. Accessed Mar. 09, 2022.
21. GDPR.EU. Complete guide to GDPR compliance. https://gdpr.eu. Accessed Mar. 15, 2022.
22. U.S. Department of Health & Human Services. Summary of the HIPAA privacy rule. https://www.hhs.gov/hipaa/for-professionals/privacy/laws-regulations/index.html. Accessed Mar. 15, 2022.
23. Laurila JK, et al. The mobile data challenge: big data for mobile computing research. Proc Work Nokia Mob Data Challenge, Conjunction with 10th Int Conf Pervasive Comput. 2012; https://doi.org/10.1016/j.pmcj.2013.07.014.
24. Blondel VD, et al. Data for development: the D4D challenge on mobile phone data. arXiv. 2012;1210(0137):1–10. Available: http://arxiv.org/abs/1210.0137
25. de Montjoye Y-A, Smoreda Z, Trinquart R, Ziemlicki C, Blondel VD. D4D-Senegal: the second Mobile phone data for development challenge. arXiv. 2014;1407(4885):1–11. Available: https://arxiv.org/pdf/1407.4885.pdf
26. De Montjoye YA, Hidalgo CA, Verleysen M, Blondel VD. Unique in the crowd: the privacy bounds of human mobility. Sci Rep. 2013;3(1376):1–5. https://doi.org/10.1038/srep01376.
27. Yu Z. Trajectory data mining: an overview. ACM Trans Intell Syst Technol. 2015;6(3):1–41.

28. Sakamanee P, Phithakkitnukoon S, Smoreda Z, Ratti C. Methods for inferring route choice of commuting trip from mobile phone network data. ISPRS Int J Geo-Information. 2020;6(5):306. https://doi.org/10.3390/ijgi9050306.

29. Wei LY, Zheng Y, Peng WC. Constructing popular routes from uncertain trajectories. In: Proceedings of the ACM SIGKDD International Conference on Knowledge Discovery and Data Mining; 2012. p. 195–203. https://doi.org/10.1145/2339530.2339562.

30. Lange R, et al. On a generic uncertainty model for position information. In: Lecture notes in computer science (including subseries lecture notes in artificial intelligence and lecture notes in bioinformatics); 2009. p. 76–87. https://doi.org/10.1007/978-3-642-04559-2_7.

31. Chow C-Y, Mokbel MF. Trajectory privacy in location-based services and data publication. ACM SIGKDD Explor Newsl. 2011;13(1):19–29. https://doi.org/10.1145/2031331.2031335.

32. Abul O, Bonchi F, Nanni M. Never walk alone: uncertainty for anonymity in moving objects databases. In: Proceedings–International Conference on Data Engineering; 2008. p. 376–85. https://doi.org/10.1109/ICDE.2008.4497446.

33. Nergiz ME, Atzori M, Saygin Y, Bariş G. Towards trajectory anonymization: a generalization-based approach. Trans Data Priv. 2009;2(1):47–75.

34. Samarati P. Protecting respondents' identities in microdata release. IEEE Trans Knowl Data Eng. 2001;13(6):1010–27. https://doi.org/10.1109/69.971193.

35. Gudmundsson J, Van Kreveld M. Computing longest duration flocks in trajectory data. In: Proceedings of the 14th Annual ACM International Symposium on Advances in Geographic Information Systems; 2006. p. 35–42. https://doi.org/10.1145/1183471.1183479.

36. Jeung H, Yiu ML, Zhou X, Jensen CS, Shen HT. Discovery of convoys in trajectory databases. Proc. VLDB Endow. 2008;1:1068–80. https://doi.org/10.14778/1453856.1453971.

37. Li Z, Ding B, Han J, Kays R. Swarm: mining relaxed temporal moving object clusters. Proc VLDB Endow. 2010;3:723–34. https://doi.org/10.14778/1920841.1920934.

38. Zheng K, Zheng Y, Yuan NJ, Shang S, Zhou X. Online discovery of gathering patterns over trajectories. IEEE Trans Knowl Data Eng. 2014;26(8):1974–88. https://doi.org/10.1109/TKDE.2013.160.

39. Sanches DE, Alvares LO, Bogorny V, Vieira MR, Kaster DS. A top-down algorithm with free distance parameter for mining top-k flock patterns. In: Lecture notes in geoinformation and cartography; 2018. p. 233–49. https://doi.org/10.1007/978-3-319-78208-9_12.

40. Akhter I, Sheikh Y, Khan S, Kanade T. Trajectory space: a dual representation for nonrigid structure from motion. IEEE Trans Pattern Anal Mach Intell. 2011;33(7):1442–56. https://doi.org/10.1109/TPAMI.2010.201.

41. Zhang T, Lu H, Li SZ. Learning semantic scene models by object classification and trajectory clustering. In: IEEE Conference on Computer Vision and Pattern Recognition (CVPR 2009); 2009. p. 1940–7. https://doi.org/10.1109/CVPRW.2009.5206809.

42. Sillito RR, Fisher RB. Semi-supervised learning for anomalous trajectory detection. In: Proceedings of the British Machine Vision Conference (BMVC 2008); 2008. p. 1035–44. https://doi.org/10.5244/C.22.103.

43. Ferreira N, Klosowski JT, Scheidegger CE, Silva CT. Vector field k-means: clustering trajectories by fitting multiple vector fields. Comput Graph Forum. 2013;32:201–10. https://doi.org/10.1111/cgf.12107.

44. Bashir FI, Khokhar AA, Schonfeld D. Object trajectory-based activity classification and recognition using hidden Markov models. IEEE Trans Image Process. 2007;16(7):1912–9. https://doi.org/10.1109/TIP.2007.898960.

45. Naftel A, Khalid S. Motion trajectory learning in the DFT-coefficient feature space. In: Proceedings of the Fourth IEEE International Conference on Computer Vision Systems (ICVS'06); 2006. p. 47–7. https://doi.org/10.1109/ICVS.2006.41.

46. Hu W, Li X, Tian G, Maybank S, Zhang Z. An incremental DPMM-based method for trajectory clustering, modeling, and retrieval. IEEE Trans Pattern Anal Mach Intell. 2013;35(5):1051–65. https://doi.org/10.1109/TPAMI.2012.188.

47. Nanni M, Pedreschi D. Time-focused clustering of trajectories of moving objects. J Intell Inf Syst. 2006;27(3):267–89. https://doi.org/10.1007/s10844-006-9953-7.
48. Xi L, Weiming H, Wei H. A coarse-to-fine strategy for vehicle motion trajectory clustering. In: Proceedings–International Conference on Pattern Recognition; 2006. p. 591–4. https://doi.org/10.1109/ICPR.2006.45.
49. Chen J, Wang R, Liu L, Song J. Clustering of trajectories based on Hausdorff distance. In: 2011 International Conference on Electronics, Communications and Control, ICECC 2011–Proceedings; 2011. p. 1940–4. https://doi.org/10.1109/ICECC.2011.6066483.
50. Liu MY, Tuzel O, Ramalingam S, Chellappa R. Entropy-rate clustering: cluster analysis via maximizing a submodular function subject to a matroid constraint. IEEE Trans Pattern Anal Mach Intell. 2014;36(1):99–112. https://doi.org/10.1109/TPAMI.2013.107.
51. Khoshaein V. Trajectory clustering using a variation of Fre'chet distance. University of Ottawa. 2013;
52. Ester M, Kriegel HP, Sander J, Xu X. A Density-Based Algorithm for Discovering Clusters in Large Spatial Databases with Noise. In: Proceedings of the Second International Conference on Knowledge Discovery and Data Mining (KDD'96); 1996. p. 226–31. https://doi.org/10.1016/B978-044452701-1.00067-3.
53. Lee JG, Han J, Whang KY. Trajectory clustering: a partition-and-group framework. In: Proceedings of the ACM SIGMOD International Conference on Management of Data; 2007. p. 593–604. https://doi.org/10.1145/1247480.1247546.
54. Lee JG, Han J, Li X, Gonzalez H. TraClass: trajectory classification using hierarchical region based and trajectory based clustering. Proc VLDB Endow. 2008;1(1):1081–94. https://doi.org/10.14778/1453856.1453972.
55. Zhou Y, Yan S, Huang TS. Detecting anomaly in videos from trajectory similarity analysis. In: Proceedings of the 2007 IEEE International Conference on Multimedia and Expo (ICME 2007); 2007. p. 1087–90. https://doi.org/10.1109/icme.2007.4284843.
56. Pelekis N, Kopanakis I, Kotsifakos EE, Frentzos E, Theodoridis Y. Clustering trajectories of moving objects in an uncertain world. In: Proceedings–IEEE International Conference on Data Mining, ICDM; 2009. p. 417–27. https://doi.org/10.1109/ICDM.2009.57.
57. Zhou X, Miao F, Ma H, Zhang H, Gong H. A trajectory regression clustering technique combining a novel fuzzy C-means clustering algorithm with the least squares method. ISPRS Int J Geo-Information. 2018;7(5):164. https://doi.org/10.3390/ijgi7050164.
58. Firdaus S, Uddin M. A survey on clustering algorithms and complexity analysis. Int J Comput Sci Issues. 2015;12:62–85.
59. Sokal RR, Michener CD. A statistical method for evaluating systematic relationships. Univ Kansas Sci Bull. 1958;38:1409–38.
60. Lamb DS, Downs J, Reader S. Space-time hierarchical clustering for identifying clusters in spatiotemporal point data. ISPRS Int J Geo-Information. 2020;9(85):1–18. https://doi.org/10.3390/ijgi9020085.
61. Zhang D, Lee K, Lee I. Hierarchical trajectory clustering for spatio-temporal periodic pattern mining. Expert Syst Appl. 2018;92(February):1–11. https://doi.org/10.1016/j.eswa.2017.09.040.
62. Zheng Y, Zhang L, Xie X, Ma W-Y. Mining interesting locations and travel sequences from GPS trajectories. In: Proceedings of the 18th international conference on World Wide Web (WWW '09); 2009. p. 791–800. https://doi.org/10.1145/1526709.1526816.
63. Von Luxburg U. A tutorial on spectral clustering. Stat Comput. 2007;17(4):32. https://doi.org/10.1007/s11222-007-9033-z.
64. Zhang Z, Huang K, Tan T, Yang P, Li J. ReD-SFA: relation discovery based slow feature analysis for trajectory clustering. In: Proceedings of the IEEE Computer Society Conference on Computer Vision and Pattern Recognition; 2016. p. 752–60. https://doi.org/10.1109/CVPR.2016.88.
65. Chen G, Lerman G. Spectral curvature clustering (SCC). Int J Comput Vis. 2009;81:317–30. https://doi.org/10.1007/s11263-008-0178-9.

66. Tuzel O, Porikli F, Meer P. Pedestrian detection via classification on Riemannian manifolds. IEEE Trans Pattern Anal Mach Intell. 2008;30(10):1713–27. https://doi.org/10.1109/TPAMI. 2008.75.

67. Ergezer H, Leblebicioğlu K. Anomaly detection and activity perception using covariance descriptor for trajectories. In: European Conference on Computer Vision–ECCV 2016 Workshops; 2016. p. 728–42. https://doi.org/10.1007/978-3-319-48881-3_51.

68. Keuper M, Andres B, Brox T. Motion trajectory segmentation via mini- mum cost multicuts. In: Proceedings of the IEEE International Conference on Computer Vision; 2015. p. 3271–9.

69. Phithakkitnukoon S, Horanont T, Witayangkurn A, Siri R, Sekimoto Y, Shibasaki R. Understanding tourist behavior using large-scale mobile sensing approach: a case study of mobile phone users in Japan. Pervasive Mob Comput. 2015;18 https://doi.org/10.1016/j.pmcj. 2014.07.003.

70. Gong L, Sato H, Yamamoto T, Miwa T, Morikawa T. Identification of activity stop locations in GPS trajectories by density-based clustering method combined with support vector machines. J Mod Transp. 2015;23:202–13. https://doi.org/10.1007/s40534-015-0079-x.

71. Xu Z, Cui G, Zhong M, Wang X. Anomalous urban mobility pattern detection based on GPS trajectories and POI data. ISPRS Int J Geo-Information. 2019;8(308):1–20. https://doi.org/10. 3390/ijgi8070308.

72. Cao H, Mamoulis N, Cheung DW. Mining frequent spatio-temporal sequential patterns. In: Proceedings of IEEE international conference on data mining (ICDM); 2005. pp. 8–15, doi: https://doi.org/10.1109/ICDM.2005.95.

73. Douglas DH, Peucker TK. Algorithms for the reduction of the number of points required to represent a digitized line or its caricature. Can Cartogr. 1973;10(2):112–22. https://doi.org/10. 1002/9780470669488.ch2.

74. Srikant R, Agrawal R. Mining sequential patterns: generalizations and performance improvements. In: Lecture Notes in Computer Science (including subseries Lecture Notes in Artificial Intelligence and Lecture Notes in Bioinformatics); 1996. p. 3–17. https://doi.org/10. 1007/bfb0014140.

75. Zaki MJ. SPADE: an efficient algorithm for mining frequent sequences. Mach Learn. 2001;42: 31–60. https://doi.org/10.1023/A:1007652502315.

76. Han J, Pei J, Mortazavi-Asl B, Chen Q, Dayal U, Hsu MC. FreeSpan: frequent pattern-projected sequential pattern mining. In: Proceeding of the Sixth ACM SIGKDD International Conference on Knowledge Discovery and Data Mining; 2000. p. 355–9.

77. Pei J, et al. PrefixSpan: mining sequential patterns efficiently by prefix-projected pattern growth. In: Proceedings–International Conference on Data Engineering; 2001. p. 215–24. https://doi.org/10.1109/icde.2001.914830.

78. Yan X, Han J, Afshar R. CloSpan: mining: closed sequential patterns in large datasets. In: Proceedings of the 3rd SIAM International Conference on Data Mining; 2003. p. 166–77. https://doi.org/10.1137/1.9781611972733.15.

79. Agrawal R, Srikant R. Fast algorithms for mining association rules. In: Proc. of 20th International Conference on Very Large Data Bases (VLDB'94); 1994, pp. 487–499.

80. Chao P, Xu Y, Hua W, Zhou X. A survey on map-matching algorithms. In: Borovica-Gajic R, Qi J, Wang W, editors. Databases theory and applications. Lecture notes in computer science, vol. 12008; 2020. p. 121–33. https://doi.org/10.1007/978-3-030-39469-1_10.

81. Giannotti F, Nanni M, Pinelli F, Pedreschi D. Trajectory pattern mining. In: Proceedings of the ACM SIGKDD International Conference on Knowledge Discovery and Data Mining; 2007. p. 63–7. https://doi.org/10.1145/1281192.1281230.

82. Ye Y, Zheng Y, Chen Y, Feng J, Xie X. Mining individual life pattern based on location history. In: Proceedings–IEEE International Conference on Mobile Data Management; 2009. p. 1–10. https://doi.org/10.1109/MDM.2009.11.

83. Xiao X, Zheng Y, Luo Q, Xie X. Inferring social ties between users with human location history. J Ambient Intell Humaniz Comput. 2014;5(1):3–19. https://doi.org/10.1007/s12652-012-0117-z.

84. Xie H, Kulik L, Tanin E. Privacy-aware traffic monitoring. IEEE Trans Intell Transp Syst. 2010;11(1):61–70. https://doi.org/10.1109/TITS.2009.2028872.

85. Song R, Sun W, Zheng B, Zheng Y. PRESS: a novel framework of trajectory compression in road networks. In: Proceedings of the VLDB Endowment; 2014. p. 661–72. https://doi.org/10.14778/2732939.2732940.

86. Wang Y, Zheng Y, Xue Y. Travel time estimation of a path using sparse trajectories. In: Proceedings of the ACM SIGKDD International Conference on Knowledge Discovery and Data Mining; 2014. p. 25–34. https://doi.org/10.1145/2623330.2623656.

Printed in the United States
by Baker & Taylor Publisher Services